Battery Management in Electric Vehicles: Current Status and Future Trends: 2nd Edition

Battery Management in Electric Vehicles: Current Status and Future Trends: 2nd Edition

Guest Editor

Prodip K. Das

MDPI

Basel • Beijing • Wuhan • Barcelona • Belgrade • Novi Sad • Cluj • Manchester

Guest Editor
Prodip K. Das
School of Engineering
University of Edinburgh
Edinburgh
UK

Editorial Office
MDPI AG
Grosspeteranlage 5
4052 Basel, Switzerland

This is a reprint of the Special Issue, published open access by the journal *Batteries* (ISSN 2313-0105), freely accessible at: https://www.mdpi.com/journal/batteries/special_issues/F4P4IHOBAK.

For citation purposes, cite each article independently as indicated on the article page online and as indicated below:

Lastname, A.A.; Lastname, B.B. Article Title. *Journal Name* **Year**, *Volume Number*, Page Range.

ISBN 978-3-7258-6332-7 (Hbk)
ISBN 978-3-7258-6333-4 (PDF)
https://doi.org/10.3390/books978-3-7258-6333-4

Contents

About the Editor

Prodip K. Das

Prodip K. Das is an Associate Professor in Hydrogen Energy Systems at the University of Edinburgh. He earned his BSc in Mechanical Engineering with distinction from the Bangladesh University of Engineering & Technology (BUET) in 1998, followed by an MSc from BUET in 2001 and another MSc from the University of Alberta in 2003. He completed his PhD in Mechanical Engineering at the University of Waterloo in 2010, specializing in hydrogen fuel cells under the guidance of Prof. Xianguo Li. Dr. Das's research interests include fuel cells, electrolyzers, Li-ion and flow batteries, transport phenomena, porous media, nanofluids, convective heat transfer, infrared thermography, and multiphysics modeling. His current work focuses on the thermofluid dynamics, design, and optimization of energy systems, including Li-ion batteries, fuel cells, and electrolyzers, as well as the safety, health, and environmental aspects of electric vehicle batteries. He has authored over 130 peer-reviewed articles, 7 book chapters, and 6 books on topics such as fuel cells, electrolyzers, Li-ion batteries, convective heat transfer, and multiphysics modeling. He is the lead editor of the invited book *Fuel Cells for Transportation: Fundamental Principles and Applications*. His work has been cited 5,000 times, resulting in an h-index of 32. In addition to his research, Dr. Das serves as Associate Editor for *Frontiers in Energy Research* (Frontiers Media), *Journal of Electrochemical Energy Conversion and Storage* (ASME), and *Frontiers in Chemical Engineering* (Frontiers Media). He is also a member of the editorial and advisory boards for several journals, including *Batteries*, *Renewable and Sustainable Energy*, *Inventions*, and *Challenges*.

Preface

This Reprint presents nine works on the critical topic of battery management in electric vehicles, covering a wide range of developments from foundational research to cutting-edge innovations. These contributions span areas such as electro-thermal modeling, state-of-health estimation, thermal runaway analysis, aging and degradation studies in lithium batteries, and the development of modular BMS topologies. Collectively, these advancements aim to optimize battery performance, enhance efficiency, and ensure the long-term reliability of EV batteries.

As the Guest Editor, I would like to express my heartfelt thanks to all the contributors who generously shared their expertise and insights, as well as to the reviewers whose thoughtful feedback helped elevate the quality and rigor of this work. I am also deeply grateful to the editorial team at *Batteries* for their unwavering support throughout this process. Additionally, I would like to acknowledge the support from the Faraday Institution (grant numbers FIRG005 and FIRG027) through the ReLiB project.

It is my sincere hope that this Reprint will serve as a valuable resource for researchers, policymakers, and graduate students in the fields of lithium-ion batteries and electric vehicles and that it contributes meaningfully to ongoing advancements in this important area of sustainable technology.

Prodip K. Das
Guest Editor

batteries

MDPI

Article

Thermal Runaway Characteristics and Gas Analysis of $LiNi_{0.9}Co_{0.05}Mn_{0.05}O_2$ Batteries

Chao Shi [1,2], Hewu Wang [1,*], Hengjie Shen [2], Juan Wang [2,*], Cheng Li [1], Yalun Li [1], Wenqiang Xu [1,3] and Minghai Li [1]

1 State Key Laboratory of Intelligent Green Vehicle and Mobility, Tsinghua University, Beijing 100084, China; 13643687239@163.com (C.S.); licheng18@mails.tsinghua.edu.cn (C.L.); liyalun@tsinghua.edu.cn (Y.L.); wenqiangxu1124@163.com (W.X.); dlminghai8813@djtu.edu.cn (M.L.)

2 College of Locomotive and Rolling Stock Engineering, Dalian Jiaotong University, Dalian 116028, China; 13478565707@163.com

3 School of Electrical & Electronic Engineering, Harbin University of Science and Technology, Harbin 150080, China

* Correspondence: wanghw@tsinghua.edu.cn (H.W.); uyuwj@163.com (J.W.)

Abstract: Layered ternary materials with high nickel content are regarded as the most promising cathode materials for high-energy-density lithium-ion batteries, owing to their advantages of high capacity, low cost, and relatively good safety. However, as the nickel content increases in ternary layered materials, their thermal stability noticeably decreases. It is of paramount importance to explore the characteristics of thermal runaway for lithium-ion batteries. In this study, two high-nickel $LiNi_{0.9}Co_{0.05}Mn_{0.05}O_2$ batteries were laterally heated to thermal runaway in a sealed chamber filled with nitrogen to investigate the thermal characteristics and gas compositions. The temperature of the battery tabs was measured, revealing that both batteries were in a critical state of thermal runaway near 120 degrees Celsius. A quantitative analysis method was employed during the eruption process, dividing it into three stages: ultra-fast, fast, and slow; the corresponding durations for the two batteries were 3, 2, 27 s and 3, 3, 26 s. By comparing the changes in chamber pressure, it was observed that both batteries exhibited a similar continuous venting duration of 32 s. However, the pressure fluctuation ranges of the two samples were 99.5 and 68.2 $kPa \cdot m \cdot s^{-1}$. Compared to the other sample, the 211 Ah sample exhibited larger chamber pressure fluctuations and reached higher peak pressures, indicating a higher risk of explosion. In the experimental phenomenon captured by a high-speed camera, it took only 1 s for the sample to transition from the opening of the safety valve to filling the experimental chamber with smoke. The battery with higher energy density exhibited more intense eruption during thermal runaway, resulting in more severe mass loss. The mass loss of the two samples is 73% and 64.87%. The electrolyte also reacted more completely, resulting in a reduced number of measured exhaust components. The main components of gaseous ejections are CO, CO_2, H_2, C_2H_4, and CH_4. For the 211 Ah battery, the vented gases were mainly composed of CO (41.3%), CO_2 (24.8%), H_2 (21%), C_2H_4 (7.4%) and CH_4 (3.9%), while those for the other 256 Ah battery were mainly CO (30.6%), CO_2 (28.5%), H_2 (21.7%), C_2H_4 (12.4%) and CH_4 (5.8%). Comparatively, the higher-capacity battery produced more gases. The gas volumes, converted to standard conditions (0 °C, 101 kPa) and normalized, resulted in 1.985 L/Ah and 2.182 L/Ah, respectively. The results provide valuable guidance for the protection of large-capacity, high-energy-density battery systems. The quantitative analysis of the eruption process has provided assistance to fire alarm systems and firefighting strategies.

Keywords: $LiNi_{0.9}Co_{0.05}Mn_{0.05}O_2$ lithium-ion battery; thermal runaway; thermal behavior; gas analysis

1. Introduction

Lithium-ion batteries (LIBs) are widely used in portable electronic devices, electric vehicles, energy storage systems, and other fields. However, during their usage, especially under abuse conditions such as high temperatures, overcharging, over-discharging, and mechanical damage, there is a risk of thermal runaway (TR) [1–3]. This refers to the potential scenario where a lithium-ion battery (LIB) may experience a violent exothermic reaction internally, leading to a loss of control and, in severe cases, causing fires or explosions. Therefore, the study of TR characteristics and gas generation in LIBs holds significant background and importance [4,5]. It is conducive to formulating more scientifically sound safety designs and usage strategies, thereby enhancing the safety performance of LIBs [6,7].

Firstly, the TR of LIBs poses a potential threat to the environment. Large-scale applications of LIBs in electric vehicles and energy storage systems may lead to significant environmental and ecological impacts in the event of a TR incident [8]. This could result in air and water pollution, presenting a potential threat to ecosystems and human health. Therefore, conducting in-depth research on the characteristics and gas generation of tTR in LIBs is beneficial for mitigating potential harm to the environment and promoting the development of sustainable energy storage technologies.

Moreover, research on the TR of LIBs offers valuable insights for addressing the challenges of energy transition and sustainable development. LIBs, as a critical energy storage technology, find widespread applications in electric vehicles and renewable energy systems and play a pivotal role in driving the adoption of clean energy. However, the safety and reliability of LIBs are essential considerations for achieving a sustainable energy transition. Investigating the TR issues associated with LIBs can help mitigate potential safety risks, providing more reliable and secure solutions for the widespread adoption of LIBs in sustainable energy systems [9].

The triggering conditions for battery TR primarily involve three aspects: electrical abuse, thermal abuse, and mechanical abuse [10]. These three aspects share a common TR trigger: an internal short circuit, which is also a primary cause of battery failure. During TR, batteries experience a phenomenon known as capacity degradation before internal heat generation, referred to as high-temperature self-discharge [11], following this, the initial heat generation occurs due to the decomposition of the Solid Electrolyte Interphase (SEI) [12]; as a result of SEI decomposition, the negative electrode comes into direct contact with the electrolyte, initiating a reaction and forming an irregular SEI [13], this leads to a continuous reaction between the negative electrode and the electrolyte, causing the internal temperature of the battery to rise until it reaches the melting point of the separator [14], the melting of the separator induces an endothermic reaction, slowing down or even reducing the rate of temperature increase within the battery. Following the melting of the separator, an internal short circuit occurs, leading to jelly roll failure and the release of a significant amount of heat. At this point, various chemical reactions occur simultaneously, such as positive electrode decomposition and electrolyte decomposition. The electrolyte decomposition generates a substantial amount of gas [15], resulting in an increase in internal gas pressure. The battery swells, and upon reaching a certain threshold, the safety valve opens, releasing high-pressure gas [16], the expelled gas contains combustible components. In the presence of oxygen in the environment and encountering an ignition source, it can lead to a severe combustion reaction [17]. In recent research, it has been discovered that TR can occur even in the absence of internal short circuits. The positively charged electrode material undergoes a phase transition at around 250 °C, releasing reactive oxygen species that react with the negative electrode [18,19]. Therefore, the significant heat generated by the oxidation–reduction reactions of the positive and negative electrodes could potentially be a direct cause of TR, not solely due to internal short circuits [20,21].

In summary, the research background and significance of TR in LIBs primarily lie in enhancing the safety of LIBs, reducing potential environmental hazards, advancing battery technology, and promoting sustainable energy transition. An in-depth study of the mechanisms and patterns of TR in LIBs can provide a scientific basis for battery design, production, usage, and management [22]. This, in turn, enables the optimal utilization of LIBs as a crucial energy storage technology, expanding their applications across various domains and facilitating the realization of sustainable energy and sustainable development goals.

Feng X. et al. provided an overview of the mechanisms of TR in LIBs used in electric vehicles, summarizing the abusive conditions that may lead to TR [23]. N.E. Galushkin analyzed the mechanism of TR in LIBs and found that the cause of TR in LIBs is the strong exothermic reaction of atomic hydrogen accumulated in the anode graphite during the battery cycling process [24]. Shen H. et al. compared the TR characteristics of LIBs with different cathode materials and analyzed their gas generation characteristics. They found that the gas composition of high-nickel $LiNi_{0.9}Co_{0.05}Mn_{0.05}O_2$ batteries is the most hazardous compared to other systems [25]. JH Kim compared the TR characteristics between NCM and LFP batteries and found that, compared to NCM, LFP batteries exhibit a delayed time to reach the maximum temperature during TR due to their higher thermal stability [26]. Sascha K. and his team conducted TR experiments on 51 batteries in a gas-tight container and analyzed their mass loss characteristics as well as gas compositions [27]. Yuan L. et al. conducted multiple TR temperature and gas generation characteristic tests on four batteries using ARC and found that the main gas concentrations in the emitted gases depend mainly on the chemical properties of the battery [28]. Zhang Q. et al. analyzed the effect of different state-of-charges (SOCs) on the gas composition generated by the battery and found that batteries at 100% (state-of-charge) SOC generate a large amount of unsaturated hydrocarbons, which pose a higher thermal hazard [29].

High-nickel layered ternary materials, as positive electrode materials for lithium-ion batteries possess numerous advantages, such as high energy density, excellent cycling stability, outstanding power performance, high resource utilization efficiency, and environmental friendliness. For batteries, the typical hazards resulting from TR are often associated with the nickel content in the positive electrode material [30,31] and the initial SOC [32], pouch cells and cylindrical cells without safety valves pose increased risks in the event of TR [33]. A higher Ni content leads to an increase in capacity, as Ni serves as the primary redox material in the core structure; the thermal stability of $Li[Ni_{1-x-y}Co_xMn_y]O_2$ is directly related to the Ni content; Mn ensures excellent cycle life and safety and Co increases electronic conductivity [31]. For the same positive electrode material at different SOCs, as the SOC increases, the onset temperature of self-heating reactions shifts towards lower temperatures. Additionally, the self-heating rate exponentially increases, especially beyond 50% SOC, significantly affecting the thermal instability [34].

Currently, most research focuses on small-capacity batteries like the 18650 and batteries with lower energy density. There is relatively less research on high-nickel NCM 9-series and high-capacity batteries. Quantitative analysis of battery venting processes remains unclear. The research primarily concentrates on comparative analyses of TR characteristics among various battery systems or comparative analyses of different TR triggering mechanisms. This study involved two different high-nickel NCM9 0.5 0.5 lithium-ion batteries with varying capacities and energy densities. In a sealed chamber, TR was induced through lateral heating within an inert gas environment. The study monitored temperature changes at the electrode tabs during the TR process, along with variations in battery voltage and chamber pressure. High-speed cameras were used to record the TR phenomenon. It was observed that the critical temperature for TR at the electrode tabs was roughly the same for both batteries. However, the higher energy density sample exhibited significantly more intense venting during the

TR process. The study employed a quantitative analysis of venting processes to compare the venting characteristics during the TR of the two batteries. Both batteries exhibited similar venting durations, but the higher energy density battery showed significantly higher fluctuations in its venting curve. Finally, the gases vented during the TR were collected and subjected to gas composition analysis.

2. Experimental Setup

2.1. Battery Samples

The two selected samples for this experiment are both batteries with $LiNi_{0.9}Co_{0.05}Mn_{0.05}O_2$ as the cathode material. Their capacities are 211 Ah and 256 Ah, respectively, with the anode primarily composed of graphite. For ease of expression, the following will use Sample A and Sample B to represent the two batteries. In this experiment, both battery samples were subjected to three charge and discharge cycles at a rate of 1/3 C. The charge and discharge cutoff voltages are specified in Table 1. The specific parameters of the battery samples are outlined in Table 1 below:

Table 1. Battery Sample Parameters.

Cell	Sample A	Sample B
Shape	Prismatic	Prismatic
Packaging Material	Aluminum Alloy	Aluminum Alloy
Capacity (Ah)	211	256
Upper Cutoff Voltage (V)	4.25	4.35
Lower Cutoff Voltage (V)	2.8	2.5
Energy Density (Wh/kg)	342.06	325.82
Weight (g)	2621.6	3417.8
SOC	100%	100%
Anode	Graphite	Graphite
Cathode	$LiNi_{0.9}Co_{0.05}Mn_{0.05}O_2$	$LiNi_{0.9}Co_{0.05}Mn_{0.05}O_2$

2.2. Experimental Setup

A 1000 L sealed container filled with nitrogen was employed to carry out TR tests and facilitate the collection of gases generated after the TR of the samples. The container is equipped with two wires, two pressure sensors, and multiple K-type thermocouples. The two wires are used for real-time monitoring of battery voltage, the pressure sensors are used for real-time monitoring of pressure changes inside the chamber, and the thermocouples are used to monitor environmental temperature inside the chamber and surface temperature changes of the samples during TR. Sensors are connected to a high-speed data acquisition system to collect experimental data with a sampling frequency of 10 Hz. The chamber door is hydraulically rotated and fixed to achieve sealing. Vacuum and nitrogen-filling pipelines are provided to create an inert atmosphere for the experiments. An observation window is located on the side of the equipment, allowing real-time observation of the experimental situation inside the chamber. A high-speed camera is placed outside the observation window to record the moment of battery ejection. The structure of the experimental chamber is shown in Figure 1. After the experiment, gases are collected and analyzed using a gas chromatograph. The gas analysis equipment used in the experiment is the Thermo Fisher Scientific gas chromatograph Trace 1300 (Country of origin and manufacturer: Thermo Fisher, Singapore), equipped with four detectors and eight chromatographic columns.

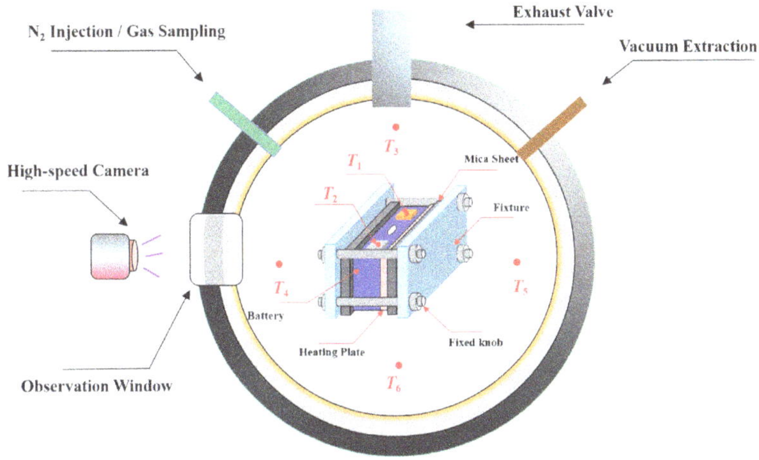

(a)

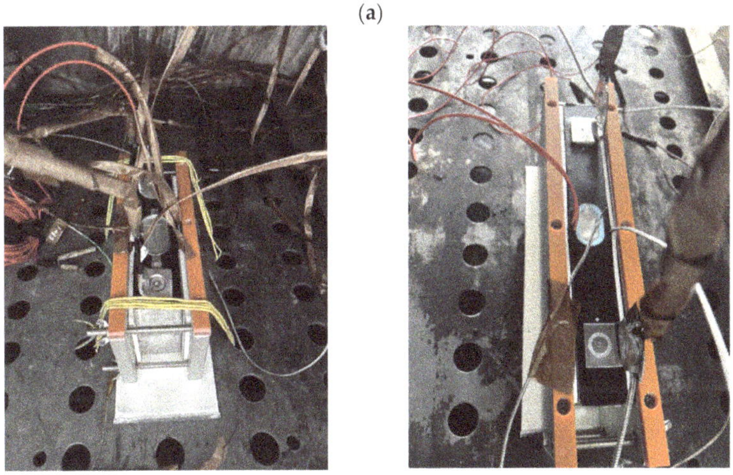

(b)

Figure 1. Experimental chamber layout. (**a**) Schematic diagram of experimental arrangement; (**b**) Experimental setup, left (Sample A) and right (Sample B).

2.3. Testing Procedure

The aim of this study was to investigate the TR characteristics and gas analysis of lithium-ion batteries with the same cathode material but different capacities. The TR was triggered by applying lateral heating to the battery using a heating plate with the same power.

The experimental procedure was as follows:

(a) The examined battery was fully charged; the voltage and weight of the battery were recorded before the experiment;

(b) Place the battery at the center of the chamber on a fixed plate. Two thermocouples (T_1, T_2) were adopted to monitor the temperature variations of the positive and negative tabs during the experiment. Another four thermocouples (T_3–T_6) were evenly fixed in the chamber in the up, down, left, and right directions to monitor the ambient temperature inside the chamber;

(c) The heating plate was positioned on one side of the battery to apply heat. Mica boards were placed on the outer side of the heating plate and the opposite side of the battery's larger surface to prevent heat loss. Clamps were used on the outermost side as fixtures to secure the setup. The wires were connected to the battery terminals to monitor voltage changes throughout the experiment;

(d) After closing the chamber door, a vacuum was drawn, N_2 was injected to restore normal pressure, and this sequence was repeated multiple times to ensure an inert gas atmosphere;

(e) A high-speed camera was placed outside the observation window to capture the eruptive state during TR;

(f) The data logger was initiated to gather data from the battery voltage, thermocouples, and pressure sensors. The heating plate was activated until the battery experienced TR, subsequently, the heating plate was deactivated.

After the experiment concluded, a period of stillness was observed until the gases inside the chamber stabilized, and the particles settled. The gases were collected from the chamber and the gas composition was analyzed using a gas chromatograph for both battery samples. Additionally, any particles and remaining battery debris were collected for weighing, and conducting an analysis of the mass loss.

3. Results and Discussion

3.1. The TR Characteristics of Batteries with Different Capacities

At elevated temperatures, the initial heat generation in the battery arises from the decomposition of the SEI membrane at the anode/electrolyte interface. This is followed by heat absorption by the separator, the heat released from the exothermic reaction at the anode/electrolyte interface, and further heat generation from the decomposition of the SEI membrane at the anode/electrolyte interface. The heat primarily originates from reactions between the anode's active material and the electrolyte, with the decomposition reaction of the cathode's active material ultimately releasing a significant amount of heat [35]. To ensure the accuracy of the chamber's environmental temperature during the experiment, the average value of the four environmental temperatures arranged was taken to represent the temperature inside the chamber. Taking the moment of TR initiation as "0" time, and considering the voltage drop as the trigger time for TR, heating by the heating plate was stopped after the TR was triggered. Figure 2 illustrates the variations in tab temperature and voltage of the two batteries, as well as the changes in environmental temperature and chamber pressure inside the experiment chamber.

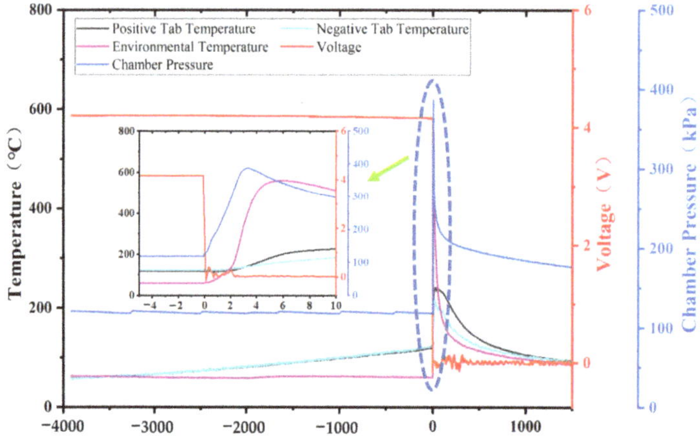

Figure 2. *Cont.*

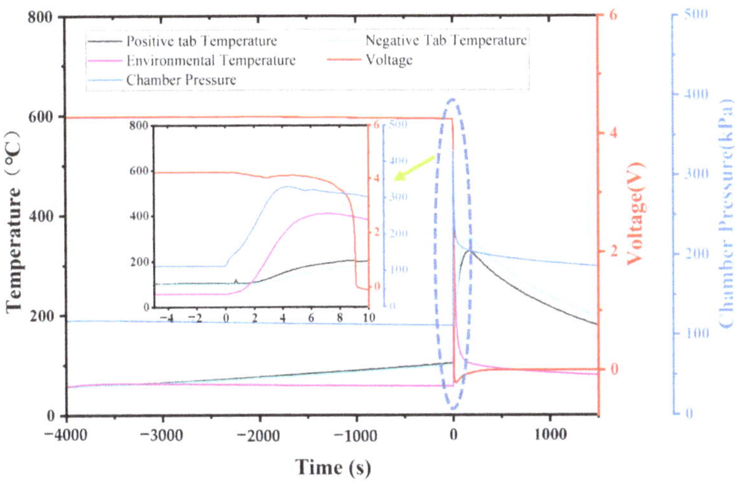

(b)

Figure 2. During the experimental process, the temperatures of the positive and negative tabs of the battery, the surface temperature of the battery, the environmental temperature, the pressure, and the voltage were monitored. (**a**) For Sample A; (**b**) For Sample B.

In the temporal sequence of reactions occurring within the battery's internal structure as temperature changes, the metastable components of the SEI film undergo exothermic decomposition between 70 and 90 degrees. Between 90 and 260 degrees, three simultaneous chemical reactions occur, including the melting of the SEI and the decomposition of various positive electrode materials, which may result in internal short circuits between electrodes. Electrolyte decomposition occurs between 200 and 300 degrees [25]. Christensen et al. conducted thermal abuse and mechanical abuse experiments on battery modules, observing trends similar to those depicted in Figure 2. However, due to differences in the triggering mechanisms between their experiment and ours, the curves exhibit less fluctuation compared to theirs [36]. The differences in the triggering mechanisms for thermal runaways result in varying characteristics of thermal runaways as well.

The initial temperature of the batteries was 30 °C, with initial voltages of 4.2 V and 4.3 V, respectively. The batteries were heated from one side, causing the battery temperature and tab temperatures to gradually rise. For Sample A, it took 3916 s to turn on the heating plate to reach the TR triggering condition. The critical temperature for TR is defined as a temperature change rate of 1 °C/s. The temperature of the positive tab rose to 119 °C, and the negative tab rose to 123 °C. The voltage stabilized at 4.16 V, and then the battery underwent TR, with the voltage rapidly dropping to 0 V. Following this, the temperatures of both tabs rose rapidly. The highest temperature inside the chamber reached 557.5 °C, and the highest pressure inside the chamber reached 386.5 kPa. For Sample B, it took 4346 s, from turning on the heating plate to reaching the TR triggering condition. The temperature of the positive tab rose to 110.9 °C, and the temperature of the negative tab rose to 108.4 °C. The voltage experienced a slight decrease when there was a sudden rise in temperature and pressure, and then TR occurred, with the voltage dropping to 0 V. The temperatures of both tabs rose rapidly. The pressure inside the chamber and the environmental temperature rose rapidly before the voltage dropped, indicating that the battery's safety valve opened before the battery failed. The reason for the difference from the phenomenon observed in sample A, where the voltage dropped before the valve opening, might be due to an initial TR occurring in a specific part of the battery core. This led to an increase in internal pressure, causing the safety valve to open. As the reaction intensified further, the entire core failed, resulting in a complete TR and the voltage dropping to 0 V. The highest temperature inside the chamber reached 409.86 °C, and the highest pressure inside the chamber was 330.5 kPa. For Sample A, the highest temperature of the positive tab was 241 °C, and the highest temperature of the negative tab was 217 °C. For Sample B, the highest temperature of the positive tab was 328 °C, and the highest temperature of the negative tab was 355.5 °C. Table 2 shows the key temperature points for two samples. Figure 3 shows the temperature changes of the positive and negative tabs during the test.

Table 2. The TR temperature data of the batteries.

Sample	T_{Pmax} (°C)	T_{Nmax} (°C)	$T_{Pcritical}$ (°C)	$T_{Ncritical}$ (°C)
A	241	217	119	123
B	328	355	110.9	108

T_{PMax}: The highest temperature during positive electrode TR. T_{NMax}: The highest temperature during negative electrode TR. $T_{Pcritical}$: The critical temperature for positive electrode TR. $T_{Ncritical}$: The critical temperature for negative electrode TR.

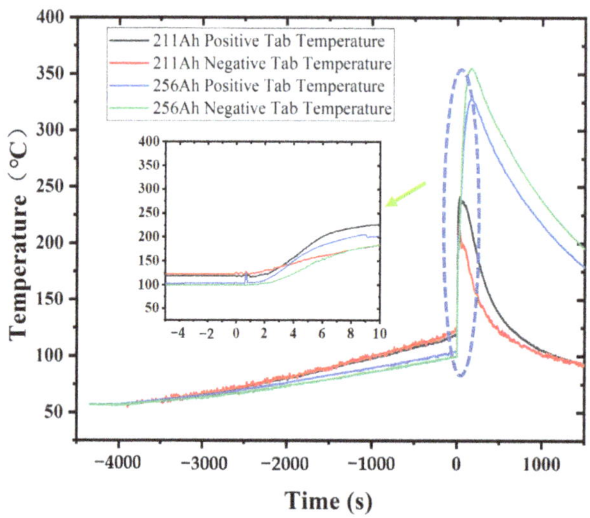

Figure 3. The temperature changes of the positive and negative tabs of the samples over time during the experimental process.

To investigate the TR spray process of these two batteries, a quantitative analysis method proposed by Zhang et al. for the spray process in LIBs was employed. The LIB spray process was divided into the ultra-fast spray stage, fast spray stage, and slow spray stage. Based on the calculation formula for the explosion index K_g, as shown in Equation (1), the normalized pressure rise rate K_{LIB} during the LIBs spray process was obtained as the spray index, as shown in Equation (2). Four typical times were defined:

1. The time corresponding to the rapid rise of K_{LIB} is defined as the start time of the spray (t_e);
2. The time corresponding to $K_{LIB,max}$ is defined as the end time (t_u) of the ultra-fast burst;
3. The time when K_{LIB} changes from positive to negative is defined as the end time (t_f) of the fast spray;
4. The time when K_{LIB} returns to the initial fluctuation state before the burst is defined as the end time (t_s) of the slow burst.

The durations corresponding to the above critical times are defined as follows:

1. The time interval from t_e to t_u is referred to as the duration of the ultra-fast burst (D_u);
2. The time interval from t_u to t_f is referred to as the duration of the fast burst (D_f);

3. The time interval from t_f to t_s is referred to as the duration of the slow burst (D_s);
4. The time interval from t_e to t_s is referred to as the overall duration of the burst (D_e) [37].

$$K_g = \left(\frac{dP}{dt}\right)_{max} V^{\frac{1}{3}} \tag{1}$$

$$K_{LIB} = \left(\frac{dP}{dt}\right) V^{\frac{1}{3}} \tag{2}$$

According to the above research method, the TR spray processes of the two batteries were compared during the experimental process, as shown in Figure 4. It can be observed that the chamber pressure follows a similar trend over time during the TR process, increasing first, then decreasing and finally stabilizing. However, the amplitude of the pressure variation for Sample A is significantly higher than that of Sample B, with D_u, D_f, D_s, D_e being 3, 2, 27, 32 s and 3, 3, 26, 32 s, respectively. The pressure characteristic parameters of the two batteries are presented in Table 3 below. Compared to Zhang's analysis of NCM6 2 2, these two batteries demonstrate shorter eruption times. This could be attributed to the higher Ni content in the samples examined in this research, intensifying the redox reactions during TR and significantly increasing the gas generation rate. Consequently, the eruption duration decreases. The higher K_{LIBmax} in both samples further supports this observation. The comparatively higher explosion risk in A, as opposed to B, could be linked to A's higher energy density.

Table 3. Pressure fluctuation parameter.

Sample	P_{max} (kPa)	$K_{LIB\ max}$ (kPa·m·s^{-1})	D_u (s)	D_f (s)	D_s (s)	D_e (s)
A	380.5	99.5	3	2	27	32
B	330.3	68.2	3	3	26	32
$Li(Ni_{0.6}Mn_{0.2}Co_{0.2})O_2$ [37]	126.3	24.9	4	4	28	36

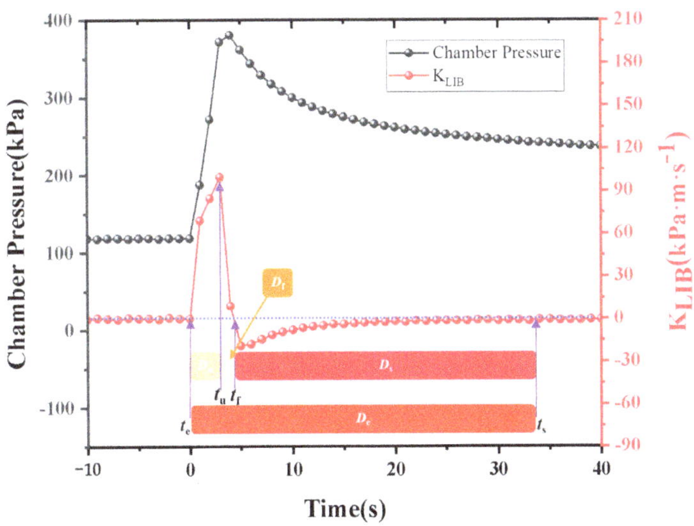

(a)

Figure 4. *Cont.*

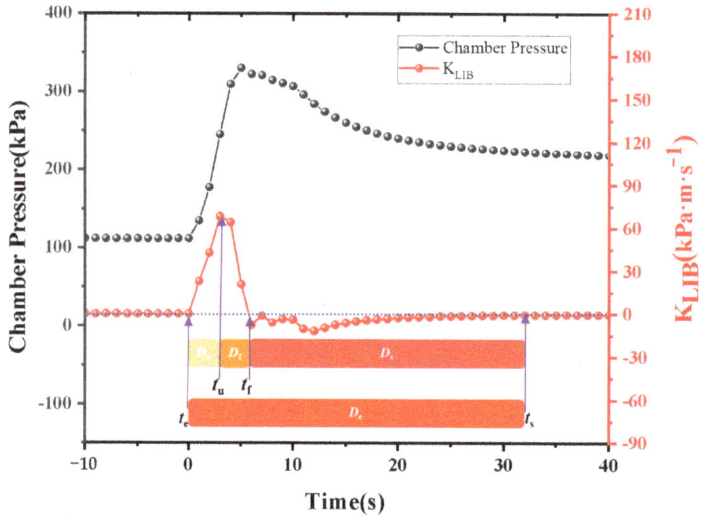

(**b**)

Figure 4. Graphs of Chamber Pressure and K_{LIB} Over Time. (**a**) For Sample A, (**b**) For Sample B.

3.2. Recording of Experimental Observations and Battery Mass Loss

The occurrence of battery rupture is primarily attributed to two factors. Firstly, explosions happen when the electrode assembly fails to maintain an opening for sufficient gas flow to reach the exhaust port. This leads to displacement of the electrode assembly, blocking the exhaust port. Secondly, when the exhaust port itself does not allow enough gas to escape, internal pressure builds up until the battery ruptures at a critical pressure [38]. In this experiment, battery rupture mainly resulted from the second cause. Wang et al. categorized the TR mass loss into three stages: (1) Packaging thermal decomposition; (2) Violent gas or smoke jetting and intense burning of ejected materials; (3) Attenuation and eventual extinction of open flames [39]. During the experiment, due to the vigorous reactions between the positive and negative electrodes and the electrolyte, a large amount of gas, particles, and electrolytes were generated. When the pressure reached a certain level, the safety valve opened, ejecting gas, particles, electrolytes, etc., resulting in battery mass loss. Figure 5 shows several images recorded during the TR process using a high-speed camera. The moment of smoke ejection from the safety valve was taken as the starting point. Within 0.06 s after the smoke ejection, sparks were observed, but at this point, smoke was still predominant. Around 0.2 s later, the ejection phenomenon was dominated by flames. It took only 1 s from the observation of smoke ejection to the entire chamber being filled with smoke within the field of view. Subsequently, the open flames gradually attenuated and eventually extinguished. We observed that the severity of TR in Sample A was higher than in Sample B. The jetting of sparks was more pronounced in Sample B. This may be attributed to the higher energy density of Sample A, leading to more intense reactions.

The weights of the two samples were measured before and after the experiments. Table 4 presents the resulting mass loss due to TR. Yang et al. compared the mass loss after TR in different systems and pointed out that NCM9 0.5 0.5 exhibited higher energy density, leading to higher mass loss. In their experiments, the mass loss was 72.89% with an energy density of 328.78 Wh/kg [40]. Sample A showed a larger mass loss after TR, indicating a more severe TR phenomenon. This was further confirmed by the recorded videos, which clearly showed more intense combustion in Sample A.

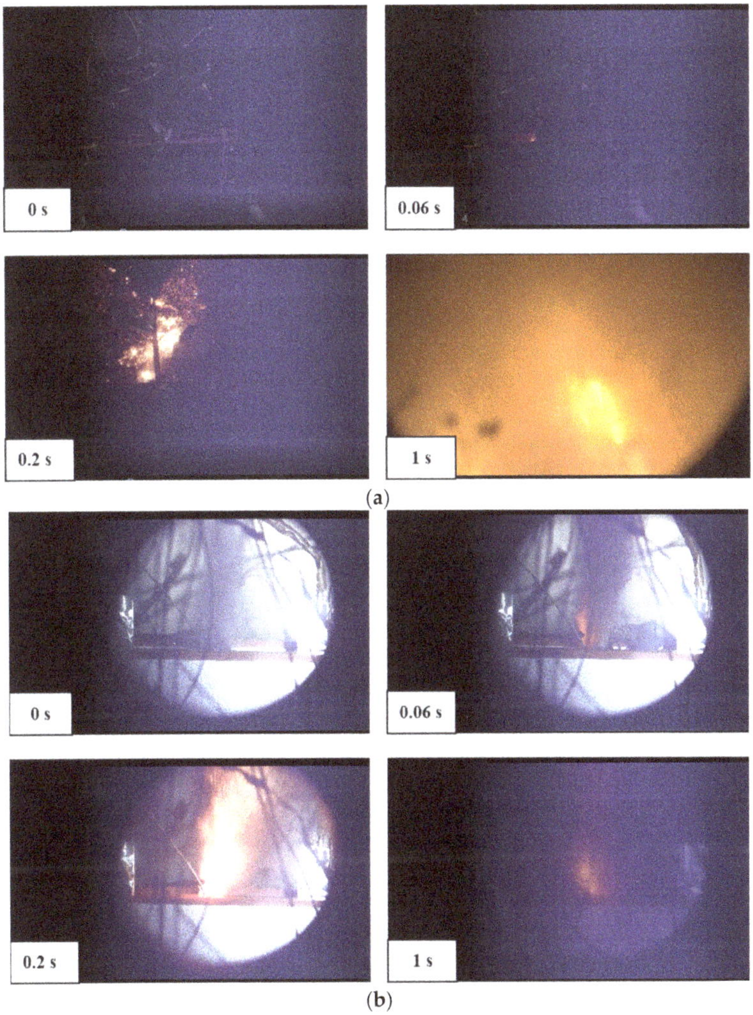

Figure 5. High-speed photography captured the eruptive state during the TR of the batteries. (**a**) Sample A; (**b**) Sample B.

Table 4. Mass Statistics.

Sample	Initial Weight (g)	Final Weight (g)	Mass Loss Rate (%)	Energy Density (Wh/Kg)
A	2621.6	707.8	73	342.06
B	3417.8	1200.6	64.87	325.82

3.3. *Gas Composition and Explosive Limits*

The TR in lithium-ion battery systems can generate hundreds of liters of high-temperature, toxic, and flammable gases. With the continuous increase in energy density and battery capacity, the exhaust from battery TR has also changed accordingly. For passenger safety in electric vehicles, it is crucial to understand the volume and composition of the exhaust gases. The experimental setup used in this study can achieve an inert gas atmosphere, filled with nitrogen, effectively preventing the reaction of gases emitted from battery TR with oxygen and other gases, ensuring the accurate measurement of the volume and composition of gases produced during TR.

The calculation of gas production during battery TR uses the ideal gas state equation, as shown in Equation (3), where n is the molar mass of the substance, P is the gas pressure, V is the gas volume, R is the ideal gas constant, and T is the ambient temperature. To ensure the accuracy of temperature measurements, multiple thermocouples are evenly distributed inside the experimental chamber, and the average value is calculated as the ambient temperature.

$$PV = nRT \tag{3}$$

$$\Delta n = \frac{P_1}{RT_1} - \frac{P_0}{RT_0} \tag{4}$$

The gas production can be calculated using Equation (4), where Δn represents the change in gas volume inside the chamber before and after the experiment, which is the volume of gas produced during battery TR. P_1 represents the pressure inside the chamber after the experiment, V is the volume of the experimental chamber (1000 L in this case), R has a value of 8.31, T_1 represents the ambient temperature inside the chamber after the experiment, P_0 represents the pressure inside the chamber before the experiment, and T_0 represents the ambient temperature inside the chamber before the experiment. To ensure accuracy in calculations, the values of P_1 and T_1 are taken from the data after the TR stabilizes to obtain the molar quantity of gas. After calculations, the gas production for Sample A is 18.35 mol, and for Sample B, it is 24.85 mol. To facilitate comparison, the gas production is converted to standard conditions, resulting in 1.958 L/Ah for Sample A and 2.182 L/Ah for Sample B. It can be observed that gas production is related to the capacity of the battery. During the thermal decomposition process, gases mainly originate from three sources: Firstly, the breakdown of SEI and NCM materials at high temperatures generates oxygen [41], and through secondary reactions with electrolytes, produces carbon dioxide. Secondly, lithium embedding into the graphite anode at high temperatures results in the release of carbon dioxide [42]. Lastly, the electrolyte decomposition at elevated temperatures produces gases such as HF, CO_2, and C_2H_4 [43]. Table 5 below shows the detected components of gases for both samples. As the experiment was conducted in an inert environment, the reactions within the chamber did not generate HF, and therefore, this component was not detected.

Table 5. Gas composition.

Sample	A	B
Major components (%)	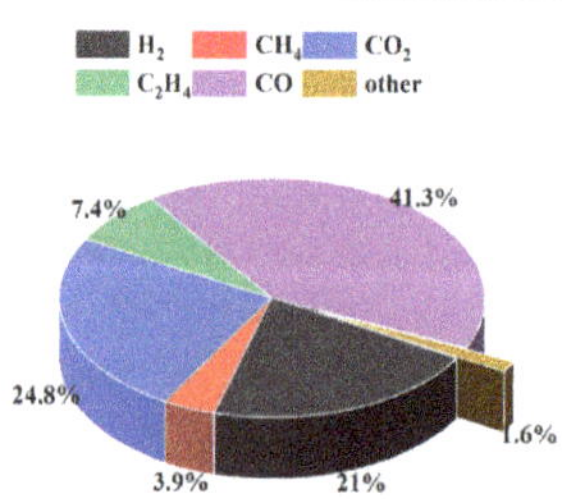	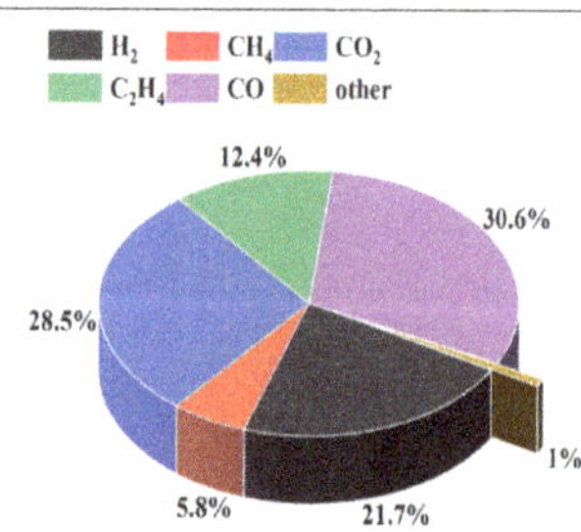
C_2H_6	0.86%	0.53%
C_3H_8	0.02%	0.04%
C_3H_6	0.19%	/
C_2H_2	0.01%	0.04%
C_4H_{10}	0.32%	0.02%
C_4H_8	/	0.05%
C_3H_4	0.15%	0.29%
C_4H_6	/	0.03%

The composition of the gases after TR is closely related to the phenomenon. When TR is accompanied by combustion, the electrolyte components in the collected gas sample will decrease [44]. The severity of combustion in Sample A is higher than that in Sample B. The electrolyte components ejected will also be more fully ignited. The detected electrolyte components of the TR gases from the two batteries are also significantly higher for the latter compared to the former.

In this section, the flammable limits of the gases generated during battery TR were also studied. Based on the gas composition measured in Table 5, the flammable upper and lower limits of the mixed gases were calculated using the Le Chatelier equation. In the equation, L_{mix} represents the flammability limit of gases generated during battery TR, L_i denotes the flammability limit of combustible component i in the battery, and x_i represents the volume percentage of component i in the mixed gas.

$$L_{mix} = \frac{1}{\sum_{i=1}^{n} \frac{x_i}{L_i}} \times 100\% \quad (5)$$

Using Equation (5) to calculate the flammable limits of the mixed gases produced during the TR of the two samples, the calculated results are shown in Table 6 below.

Table 6. The flammable limits.

Sample	UFL	LFL	Flammable Range
A	58.3	8.49	49.81
B	51.88	7.43	44.45

The safety of batteries is assessed from two aspects of gas flammability limits. On one hand, the lower flammable limit indicates how easily the gas can act as a fuel and lead to combustion or explosion; thus, the lower the flammable limit, the higher the risk. On the other hand, the flammable concentration range reflects how easily the gas in the environment can satisfy the conditions for combustion. A wider flammable concentration range implies that the gas is more likely to combust under environmental conditions.

4. Conclusions

In this study, we investigated the characteristics of terminal temperature and gas composition during the TR of batteries with the same positive electrode material NCM9 0.5 0.5 but different capacities. The experimental samples had capacities of 211 Ah (Sample A) and 256 Ah (Sample B), with the energy density of Sample A being higher than that of Sample B. The main experimental equipment was a sealed chamber, and the experimental environment was a sealed inert atmosphere. The TR was triggered by lateral heating. The main conclusions are as follows:

(1) The TR triggering temperatures for the two batteries are relatively close, with the electrode tab temperatures both around 120 °C. The maximum temperature of Sample B electrode plates is significantly higher than that of Sample A. Yet, the maximum pressure within Sample A chamber is higher than that of Sample B.

(2) Quantitative analysis of the spray process for the two lithium-ion batteries was performed using the K_{LIB} curve. The ultra-fast, fast, and slow venting stages for both batteries were observed to be 3, 2, 27 s and 3, 3, 26 s. The total venting duration for both was 32 s. However, it was evident that the fluctuation amplitude of the chamber pressure for Sample A was notably higher than that of Sample B; A poses a higher explosion risk compared to B;

(3) In the experimental phenomenon captured by the high-speed camera, it took only 1 s for the sample to transition from the opening of the safety valve to filling the experimental chamber with smoke. The battery with higher energy density in the two exhibits a higher degree of severity in TR, with more pronounced phenomena of

jetting flames and a significantly increased mass loss rate. The mass loss rates for A and B were 73% and 64.87%;

(4) The gas production of the two batteries under standard conditions is 1.985 L/Ah for one and 2.182 L/Ah for the other. With gas production increasing with battery capacity. The main gas components for both batteries, listed in descending order, are CO, CO_2, H_2, C_2H_4, and CH_4. The total amount of gas components measured in Sample A is lower than that in Sample B.

5. The Limitation of the Study and the Future Work

This study only conducted comparisons with the same positive electrode materials, focusing solely on the differences in capacity and energy density. Subsequent additions should include comparisons of TR temperatures and gas production characteristics for different positive electrode materials, capacities, and application types. The aim is to provide assistance for the application of batteries in various scenarios and for thermal hazard protection.

Author Contributions: Conceptualization, C.S. and C.L.; methodology, C.S.; software, C.S. and H.S.; validation, H.W., J.W. and M.L.; formal analysis, C.S.; investigation, C.S.; resources, H.W.; data curation, H.S.; writing—original draft preparation, C.S.; writing—review and editing, H.W., C.L., Y.L. and M.L.; visualization, C.S., W.X. and H.S.; supervision, J.W. All authors have read and agreed to the published version of the manuscript.

Funding: Sponsored by Tsinghua-Toyota Joint Research Fund.

Data Availability Statement: The data are available through appropriate requests.

Conflicts of Interest: The authors declare no conflicts of interest.

Nomenclature

Symbols	Explanation
LIB	lithium-ion battery
LIBs	lithium-ion batteries
SEI	Solid Electrolyte Interphase
SOC	state-of-charge
SOCs	state-of-charges
NCM	$LiNi_xCo_yMn_zO_2$
UFL	Upper flammable limit
LFL	Lower flammable limit
TR	Thermal runaway
T_{PMax}	The highest temperature during positive electrode TR
T_{NMax}	The highest temperature during negative electrode TR
$T_{Pcritical}$	The critical temperature for positive electrode TR
$T_{Ncritical}$	The critical temperature for negative electrode TR
P_1	the pressure inside the chamber after the experiment
V	the volume of the experimental chamber
T_1	the ambient temperature inside the chamber after the experiment
P_0	the pressure inside the chamber before the experiment
T_0	the ambient temperature inside the chamber before the experiment
L_{mix}	the flammability limit of gases generated during battery TR
L_i	the flammability limit of combustible component *i* in the battery
x_i	the volume percentage of component *i* in the mixed gas

References

1. Wang, X.; Wei, X.; Zhu, J.; Dai, H.; Zheng, Y.; Xu, X.; Chen, Q. A review of modeling, acquisition, and application of lithium-ion battery impedance for onboard battery management. *eTransportation* **2021**, *7*, 100093. [CrossRef]
2. Lai, X.; Chen, Q.; Tang, X.; Zhou, Y.; Gao, F.; Guo, Y.; Bhagat, R.; Zheng, Y. Critical review of life cycle assessment of lithium-ion batteries for electric vehicles: A lifespan perspective. *eTransportation* **2022**, *12*, 100169. [CrossRef]

3. Cai, T.; Valecha, P.; Tran, V.; Engle, B.; Stefanopoulou, A.; Siegel, J. Detection of Li-ion battery failure and venting with Carbon Dioxide sensors. *eTransportation* **2021**, *7*, 100100. [CrossRef]
4. Piao, N.; Gao, X.; Yang, H.; Guo, Z.; Hu, G.; Cheng, H.-M.; Li, F. Challenges and development of lithium-ion batteries for low temperature environments. *eTransportation* **2022**, *11*, 100145. [CrossRef]
5. Xu, W.; Wu, X.; Li, Y.; Wang, H.; Lu, L.; Ouyang, M. A comprehensive review of DC arc faults and their mechanisms, detection, early warning strategies, and protection in battery systems. *Renew. Sustain. Energy Rev.* **2023**, *186*, 113674. [CrossRef]
6. Faraji, H.; Yıldız, Ç.; Arshad, A.; Arıcı, M.; Choukairy, K.; El Alami, M. Passive thermal management strategy for cooling multiple portable electronic components: Hybrid nanoparticles enhanced phase change materials as an innovative solution. *J. Energy Storage* **2023**, *70*, 108087. [CrossRef]
7. Faraji, H.; Teggar, M.; Arshad, A.; Arıcı, M.; Mehdi Berra, E.; Choukairy, K. Lattice Boltzmann simulation of natural convection heat transfer phenomenon for thermal management of multiple electronic components. *Therm. Sci. Eng. Prog.* **2023**, *45*, 102126. [CrossRef]
8. Peiyan, Q.I.; Jie, Z.M.; Da, J.; Kai, Y.; Jianling, L.; Yilin, L.; Fei, G.; Hao, L. Combustion characteristics of lithium–iron–phosphate batteries with different combustion states. *eTransportation* **2022**, *11*, 100148. [CrossRef]
9. Wang, Q.; Mao, B.; Stoliarov, S.I.; Sun, J. A review of lithium ion battery failure mechanisms and fire prevention strategies. *Prog. Energy Combust. Sci.* **2019**, *73*, 95–131. [CrossRef]
10. Wen, J.; Yu, Y.; Chen, C. A Review on Lithium-Ion Batteries Safety Issues: Existing Problems and Possible Solutions. *Mater. Express* **2012**, *2*, 197–212. [CrossRef]
11. Ishikawa, H.; Mendoza, O.; Sone, Y.; Umeda, M. Study of thermal deterioration of lithium-ion secondary cell using an accelerated rate calorimeter (ARC) and AC impedance method. *J. Power Sources* **2012**, *198*, 236–242. [CrossRef]
12. Spotnitz, R.; Franklin, J. Abuse behavior of high-power, lithium-ion cells. *J. Power Sources* **2003**, *113*, 81–100. [CrossRef]
13. Zhou, M.; Zhao, L.; Okada, S.; Yamaki, J.I. Quantitative Studies on the Influence of LiPF_6 on the Thermal Stability of Graphite with Electrolyte. *J. Electrochem. Soc.* **2012**, *159*, A44. [CrossRef]
14. Orendorff, C.J. The Role of Separators in Lithium-Ion Cell Safety. *Electrochem. Soc. Interface* **2012**, *21*, 61. [CrossRef]
15. Botte, G.G.; White, R.E.; Zhang, Z. Thermal stability of LiPF 6 –EC:EMC electrolyte for lithium ion batteries. *J. Power Sources* **2001**, *97*, 570–575. [CrossRef]
16. Harris, S.J.; Timmons, A.; Pitz, W.J. A combustion chemistry analysis of carbonate solvents used in Li-ion batteries. *J. Power Sources* **2009**, *193*, 855–858. [CrossRef]
17. Liu, X.; Stoliarov, S.I.; Denlinger, M.; Masias, A.; Snyder, K. Comprehensive calorimetry of the thermally-induced failure of a lithium ion battery. *J. Power Sources* **2015**, *280*, 516–525. [CrossRef]
18. Lamb, J.; Orendorff, C.J.; Steele, L.A.M.; Spangler, S.W. Failure propagation in multi-cell lithium ion batteries. *J. Power Sources* **2015**, *283*, 517–523. [CrossRef]
19. Mohammadian, S.K.; Zhang, Y. Improving wettability and preventing Li-ion batteries from thermal runaway using microchannels. *Int. J. Heat Mass Transf.* **2018**, *118*, 911–918. [CrossRef]
20. Dongxu, O.; Jiahao, L.; Mingyi, C.; Jingwen, W.; Jian, W. An Experimental Study on the Thermal Failure Propagation in Lithium-Ion Battery Pack. *J. Electrochem. Soc.* **2018**, *165*, A2184–A2193.
21. Golubkov, A.W.; Scheikl, S.; Planteu, R.; Voitic, G.; Hacker, V. Thermal runaway of commercial 18650 Li-ion batteries with LFP and NCA cathodes—Impact of state of charge and overcharge. *RSC Adv.* **2015**, *5*, 57171–57186. [CrossRef]
22. Spinner, N.S.; Field, C.R.; Hammond, M.H.; Williams, B.A.; Myers, K.M.; Lubrano, A.L.; Rose-Pehrsson, S.L.; Tuttle, S.G. Physical and chemical analysis of lithium-ion battery cell-to-cell failure events inside custom fire chamber. *J. Power Sources* **2015**, *279*, 713–721. [CrossRef]
23. Feng, X.; Ouyang, M.; Liu, X.; Lu, L.; Xia, Y.; He, X. Thermal runaway mechanism of lithium ion battery for electric vehicles: A review. *Energy Storage Mater.* **2018**, *10*, 246–267. [CrossRef]
24. Galushkin, N.E.; Yazvinskaya, N.N.; Galushkin, D.N. Mechanism of Thermal Runaway in Lithium-Ion Cells. *J. Electrochem. Soc.* **2018**, *165*, A1303. [CrossRef]
25. Shen, H.; Wang, H.; Li, M.; Li, C.; Zhang, Y.; Li, Y.; Yang, X.; Feng, X.; Ouyang, M. Thermal Runaway Characteristics and Gas Composition Analysis of Lithium-Ion Batteries with Different LFP and NCM Cathode Materials under Inert Atmosphere. *Electronics* **2023**, *12*, 1603. [CrossRef]
26. Kim, J.H.; Kwak, E.G.; Jeong, J.; Oh, K.Y. Versatile multiphysics model for thermal runaway estimation of a lithium-ion battery. *Int. J. Energy Res.* **2022**, *46*, 16550–16575. [CrossRef]
27. Koch, S.; Fill, A.; Birke, K.P. Comprehensive gas analysis on large scale automotive lithium-ion cells in thermal runaway. *J. Power Sources* **2018**, *398*, 106–112. [CrossRef]
28. Yuan, L.; Dubaniewicz, T.; Zlochower, I.; Thomas, R.; Rayyan, N. Experimental study on thermal runaway and vented gases of lithium-ion cells. *Process Saf. Environ. Prot.* **2020**, *144*, 186–192. [CrossRef]
29. Zhang, Q.; Niu, J.; Zhao, Z.; Wang, Q. Research on the effect of thermal runaway gas components and explosion limits of lithium-ion batteries under different charge states. *J. Energy Storage* **2022**, *45*, 103759. [CrossRef]
30. Wang, H.; Du, Z.; Rui, X.; Wang, S.; Jin, C.; He, L.; Zhang, F.; Wang, Q.; Feng, X. A comparative analysis on thermal runaway behavior of Li (Ni Co Mn) O_2 battery with different nickel contents at cell and module level. *J. Hazard. Mater.* **2020**, *393*, 122361. [CrossRef]

31. Noh, H.J.; Youn, S.; Yoon, C.S.; Sun, Y.K. Comparison of the structural and electrochemical properties of layered $Li[Ni_xCo_yMn_z]O_2$ (x = 1/3, 0.5, 0.6, 0.7, 0.8 and 0.85) cathode material for lithium-ion batteries. *J. Power Sources* **2013**, *233*, 121–130. [CrossRef]
32. Zheng, Y.; Gao, W.; Ouyang, M.; Lu, L.; Zhou, L.; Han, X. State-of-charge inconsistency estimation of lithium-ion battery pack using mean-difference model and extended Kalman filter. *J. Power Sources* **2018**, *383*, 50–58. [CrossRef]
33. Huang, L.; Zhang, Z.; Wang, Z.; Zhang, L.; Zhu, X.; Dorrell, D.D. Thermal runaway behavior during overcharge for large-format Lithium-ion batteries with different packaging patterns. *J. Energy Storage* **2019**, *25*, 100811.
34. Jindal, P.; Bhattacharya, J. Review—Understanding the Thermal Runaway Behavior of Li-Ion Batteries through Experimental Techniques. *J. Electrochem. Soc.* **2019**, *166*, A2165–A2193. [CrossRef]
35. He, T.; Zhang, T.; Gadkari, S.; Wang, Z.; Mao, N.; Cai, Q. An investigation on thermal runaway behaviour of a cylindrical lithium-ion battery under different states of charge based on thermal tests and a three-dimensional thermal runaway model. *J. Clean. Prod.* **2023**, *388*, 135980. [CrossRef]
36. Christensen, P.; Milojevic, Z.; Wise, M.; Ahmeid, M.; Attidekou, P.; Mrozik, W.; Dickmann, N.; Restuccia, F.; Lambert, S.; Das, P. Thermal and mechanical abuse of electric vehicle pouch cell modules. *Appl. Therm. Eng.* **2021**, *189*, 116623. [CrossRef]
37. Zhang, Y.; Wang, H.; Li, W.; Li, C.; Ouyang, M. Quantitative analysis of eruption process of abused prismatic Ni-rich automotive batteries based on in-chamber pressure. *J. Energy Storage* **2020**, *31*, 101617. [CrossRef]
38. Finegan, D.P.; Darcy, E.; Keyser, M.; Tjaden, B.; Heenan, T.M.; Jervis, R.; Bailey, J.J.; Vo, N.T.; Magdysyuk, O.V.; Drakopoulos, M.J.A.S. Thermal Runaway: Identifying the Cause of Rupture of Li-Ion Batteries during Thermal Runaway. *Adv. Sci.* **2018**, *5*, 1870003. [CrossRef]
39. Wang, Z.; Yang, H.; Li, Y.; Wang, G.; Wang, J. Thermal runaway and fire behaviors of large-scale lithium ion batteries with different heating methods. *J. Hazard. Mater.* **2019**, *379*, 120730. [CrossRef]
40. Yang, X.; Wang, H.; Li, M.; Li, Y.; Li, C.; Zhang, Y.; Chen, S.; Shen, H.; Qian, F.; Feng, X.; et al. Experimental Study on Thermal Runaway Behavior of Lithium-Ion Battery and Analysis of Combustible Limit of Gas Production. *Batteris* **2022**, *8*, 250. [CrossRef]
41. Yang, H.; Shen, X.D. Dynamic TGA–FTIR studies on the thermal stability of lithium/graphite with electrolyte in lithium-ion cell. *J. Power Sources* **2007**, *167*, 515–519. [CrossRef]
42. Garg, M.; Tanim, T.R.; Rahn, C.D.; Bryngelsson, H.; Legnedahl, N. Elevated temperature for life extension of lithium ion power cells. *Energy* **2018**, *159*, 71–723. [CrossRef]
43. Galushkin, N.E.; Yazvinskaya, N.N.; Galushkin, D.N. Mechanism of Gases Generation during Lithium-Ion Batteries Cycling. *J. Electrochem. Soc.* **2019**, *166*, A897–A908. [CrossRef]
44. Zhang, Q.; Liu, T.; Wang, Q. Experimental study on the influence of different heating methods on thermal runaway of lithium-ion battery. *J. Energy Storage* **2021**, *42*, 103063. [CrossRef]

batteries

MDPI

Article

Sustainable Management of Rechargeable Batteries Used in Electric Vehicles

Jay Meegoda [1,*], Ghadi Charbel [1] and Daniel Watts [2]

1 Civil and Environmental Engineering, New Jersey Institute of Technology, Newark, NJ 07102, USA; gc383@njit.edu
2 Biology Department, Indiana University—Bloomington, Bloomington, IN 47405, USA; dajwatts@indiana.edu
* Correspondence: meegoda@njit.edu; Tel.: +1-973-596-2464

Abstract: A Life Cycle Assessment (LCA) quantifies the environmental impacts during the life of a product from cradle to grave. It evaluates energy use, material flow, and emissions at each stage of life. This report addresses the challenges and potential solutions related to the surge in electric vehicle (EV) batteries in the United States amidst the EV market's exponential growth. It focuses on the environmental and economic implications of disposal as well as the recycling of lithium-ion batteries (LIBs). With millions of EVs sold in the past decade, this research highlights the necessity of efficient recycling methods to mitigate environmental damage from battery production and disposal. Utilizing a Life Cycle Assessment (LCA) and Life Cycle Cost Assessment (LCCA), this research compares emissions and costs between new and recycled batteries by employing software tools such as SimaPro V7 and GREET V2. The findings indicate that recycling batteries produces a significantly lower environmental impact than manufacturing new units from new materials and is economically viable as well. This research also emphasizes the importance of preparing for the upcoming influx of used EV batteries and provides suggestions for future research to optimize the disposal and recycling of EV batteries.

Keywords: Life Cycle Analysis; recycling; electric vehicles; Life Cycle Cost Analysis

1. Introduction

A Life Cycle Assessment (LCA) is a methodology for assessing the environmental impacts associated with all the stages of life of a product from raw material extraction through materials processing, manufacture, distribution, use, repair and maintenance, as well as eventual disposal or recycling. An LCA aims to identify and quantify the energy and materials used and wastes released to the environment to assess their impacts on the environment, and it aims to identify and evaluate opportunities for environmental improvements. The assessment includes the entire life cycle of the product, process, or service from cradle to grave [1]

In the United States, the market for electric vehicles (EVs), fanned by initiatives to address climate change and global warming, reached a value of USD 49.1 billion in 2022. Forecasts suggest that it will expand to USD 215.7 billion by 2032 with a compound annual growth rate of 15.5% [2,3]. This surge in popularity of EVs has amplified the demand for lithium-ion batteries (LIBs), resulting in increased prices for essential raw materials of batteries such as cobalt, lithium, copper, and nickel. The prices of these raw materials have seen a dramatic increase; for example, the price of lithium increased by four to five times in 2021 and almost doubled again from January 2022 to January 2023 [4]. Such heightened demand has led to environmental and humanitarian concerns at numerous mining sites, which are likely to increase proportionally with demand unless they are addressed.

Managing the surge of used EV lithium-ion batteries presents a significant challenge for the U.S. as well as the rest of the world. The intricate composition of electric car batteries includes elements like lithium, nickel, cobalt, copper, and graphite carbon, some of which are increasingly difficult to acquire economically. The scarcity and mining intensity of these elements, coupled with pollution-heavy extraction processes, pose serious environmental threats [5]. The improper disposal of LIBs creates the risk of releasing toxic chemicals into soil and water, leading to the long-term contamination of natural resources. Such actions can result in a wide array of health issues, including damage to internal organs, respiratory problems, birth defects, and heart disease.

The ramifications of incorrect battery disposal extend beyond local pollution by contributing to global environmental issues and probable illegal e-waste dumping. With the growing global popularity of electric vehicles, improper disposal is becoming increasingly problematic, elevating pollution levels and toxic waste in various countries. This, in turn, affects water resources and public health while causing substantial harm to diverse ecosystems and wildlife [6]. Given that over 2.5 million batteries for electrical and plug-in vehicles have been sold in the past 12 years [7], efficient recycling solutions are required. The U.S. government is funding recycling projects, with the Environmental Protection Agency investing over USD 100 million in such initiatives and the Bipartisan Infrastructure Law allocating USD 275 million between 2022 and 2026 as part of the Save Our Seas 2.0 Act [8]. This study aims to assess the environmental impacts and the effectiveness of these recycling measures within the electric vehicle sector.

2. Literature Review

Global climate change has become one of the primary problems facing the planet. In response to this threat, investments have been made to facilitate the transition from fossil fuels to renewable energy as a primary power source for vehicles. One of the manifestations of these investments is the surge in numbers of electric vehicles (EVs). As EVs become more accessible and their prices continue decreasing, their adoption is expected to grow significantly, with EVs predicted to make up 10% of all vehicle sales by 2025 and up to 30% by 2030 [9]. However, this increase in EV adoption raises a consequential concern: an optimal strategy is needed for managing EV batteries when they reach their end-of-life stage. This research performed a comprehensive review of the recent scholarly literature to ascertain the most effective way of dealing with the critical issue of EV battery disposal from a Life Cycle Assessment perspective.

As stated by Bobba et al. [10], Fan et al. [11], Quan et al. [12], and Gains et al. [13], once batteries degrade to an 80% charging capacity, they are not effectively usable in EVs; nonetheless, those batteries still have significant capacity within them, making them suitable for reuse in other applications. In fact, Picatoste et al. [14] argue that, given the high cost of manufacturing EV batteries and the impact of their production on the environment, it is imperative to maximize the useful lifespans of these batteries by repurposing them in different scenarios because the energy needed to recycle batteries directly after their initial use in EVs is greater than the environmental benefits resulting from the recycling effort [15,16]. While Kotak et al. [15] contend that when it comes to EV batteries, there are countless possibilities for second-life applications, the majority of the literature identified for this research focuses on repurposing batteries within an Energy Storage System (ESS). Bobba et al. [10] highlight that repurposed EV batteries show promising environmental benefits, especially when they are employed in place of new storage batteries in order to support the self-sustaining energy needs of stand-alone photovoltaic (PV) installation in houses. The environmental advantages are more substantial in regions with a less green energy mix, such as substituting a diesel generator with grid-connected renewable energy. Nonetheless, one problem that arises when building an ESS is the multitude of battery chemistries, different degrees of usage, and the compositions available in the market, making it difficult to build a truly interconnected system [8]. The two most common battery chemistries prevalent within EVs are lithium iron phosphate (LFP) batteries and lithium nickel cobalt manganese oxide (NCM) batteries [11]. To remedy this issue, Kotak et al. [15] propose constructing an ESS with individual cell control, which would allow for the combination of different cell chemistries while maximizing performance. Nevertheless, such a structure would come with a significantly higher cost. Once the battery reaches 60% of its initial charging capacity, the battery must be disposed of as it can no longer fulfill the system requirements for any potential second-use scenario. [10]. At this stage, recycling becomes imperative in order to maximize the environmental benefits.

As reported by Fan et al. [11], recycling methods for lithium-ion batteries commonly include hydrometallurgy, pyrometallurgy, and direct recycling. They found that when comparing recycling methods, hydrometallurgical recycling lags slightly behind pyrometallurgical recycling and direct physical recycling in terms of environmental benefits, possibly due to the generation of a large amount of acidic wastewater during the hydrometallurgical process. Also, among these methods, recycling NCM batteries was found to offer better environmental benefits. Conversely, in their study, Quan et al. [12] dispute that hydrometallurgy consistently outperformed direct physical recycling and pyrometallurgy in terms of environmental benefits. In fact, their study showcased that pyrometallurgy was less effective in recycling steel, copper, and aluminum from NCM batteries. Similarly, Marchese et al. [17] present a comprehensive examination of the hydrometallurgical process, advocating the utilization of organic acids such as citric acid, oxalic acid, maleic acid, or combinations thereof to foster a more sustainable approach to metal extraction. Their analysis suggests that this method not only aligns with environmental and economic sustainability, but also circumvents the need for costly post-processing of wastewater that is typically associated with hydrometallurgical extraction techniques. Furthermore, the use of organic acids is credited with simplifying the management of the leaching solution, reducing energy consumption, and diminishing carbon dioxide emissions, thereby contributing to a reduction in the environmental footprint of metal extraction.

On the other hand, Rosenberg et al. [8] and Picatoste et al. [14] state that using both recycling technologies, pyrometallurgy and hydrometallurgy, for the entire volume of end-of-life batteries has the potential to achieve the lowest overall environmental impacts within defined limits. Finally, in their study, Zanoletti et al. [18] explore various methods for recycling lithium-ion batteries (LIBs), identifying hydrometallurgy and solvo-metallurgy as standout approaches due to their unique benefits and challenges. Hydrometallurgy is praised for its efficiency in energy use and its ability to purify metals to a high degree, effectively reducing the environmental impact. On the other hand, solvo-metallurgy, an innovative method using non-aqueous solvents like ionic liquids and deep eutectic solvents (DESs), seeks to overcome the shortcomings of hydrometallurgy by reducing the generation of wastewater and facilitating the thorough dissolution of cathode materials at reduced temperatures. This cutting-edge technique is noted for its potential to significantly improve the sustainability and effectiveness of the recycling process. Solvo-metallurgy, with its pioneering strategies and dedication to environmental conservation, has the potential to revolutionize the standards for efficient and eco-friendly battery recycling. However, its scalability and industrial application are currently limited in contrast to pyrometallurgy and hydrometallurgy, which are already widely implemented at the industrial level.

In summary, the above review highlights that while batteries may become unsuitable for EV use when they reach around 80% recharge capacity, they still retain significant potential for reuse, which is crucial given the environmental costs associated with the full scope of battery manufacturing. Repurposing these batteries, especially in Energy Storage Systems (ESSs) for self-sustaining energy needs, presents promising environmental benefits, particularly in regions with a lower availability of green energy. However, challenges exist in building interconnected ESSs due to the diversity of battery chemistries and their usage and compositions. However, it is important to note that none of the studies mentioned in this manuscript have been conducted in the United States, and given that the electricity infrastructure within each country plays a significant role in the LCA comparisons for EVs, such a study was performed and is reported here in order to clarify the true impact that those batteries have on the environment in the U.S.

3. Environmental Impact Categories

To fully understand the total environmental impact associated with EV battery recycling, standard environmental impact categories were used to quantify the adverse environmental and health impacts. Among over 30 environmental impacts considered for inclusion, 19 were selected as being the most relevant for this study: Global Warming Potential (GWP), Acidification Potential (AP), Cumulative Energy Demand (CED), Ozone Depletion Potential (ODP), Particulate Matter Formation (PMF), Abiotic Depletion Potential (ADP), Photochemical Ozone Creation Potential (POCP), Freshwater Ecotoxicity Potential (FETP), Human Toxicity Non-Carcinogenic (HTnc), Human Toxicity Carcinogenic (HTc), Eutrophication Potential for Terrestrial Situations (EPt), Eutrophication Potential for Marine Ecosystems (EPm), Water Depletion, Land Use and Land Change, Biodiversity Loss, Noise Pollution, Soil Quality Degradation, Thermal Pollution, and Groundwater Contamination. A brief description of each selected impact category is provided below:

1. **Global Warming Potential (GWP):** Global Warming Potential (GWP) is a key metric in a Life Cycle Assessment (LCA) for evaluating the impact of greenhouse gases (GHGs) on global warming. It compares the radiative forcing effect—meaning the change in the Earth's energy balance—of different GHGs to that of carbon dioxide (CO_2), the reference gas. The GWP is calculated over different timeframes, typically 20, 100, and 500 years, to account for the varying lifespans and immediate impacts of different gases. Shorter time horizons emphasize the effects of gases like methane (CH4), which are short-lived but initially highly potent, whereas longer horizons focus on gases like CO_2 that persist longer in the atmosphere. The GWP of CO_2 is set as 1 across all timeframes, and other gases are rated based on how their warming effects compare to those of CO_2. This takes into account factors like the gas's ability to absorb and emit infrared radiation, its atmospheric lifespan, and its concentration.
2. **Acidification Potential (AP):** Acidification Potential (AP) is an important category in a Life Cycle Assessment (LCA) that evaluates the potential of emissions to cause acidification in the environment. This process involves pollutants being emitted into the atmosphere that transform chemically and return to the Earth's surface as "acid rain" or other acidic substances or materials that can be converted by natural processes into acidic substances. This can have harmful effects on soil, water, ecosystems, and human health. Key pollutants contributing to acidification include sulfur dioxide (SO_2), nitrogen oxides (NOx), ammonia (NH_3), and volatile organic compounds (VOCs), which can arise from human activities as well as natural sources.
3. **Cumulative Energy Demand (CED):** Cumulative Energy Demand (CED) is a category in a Life Cycle Assessment (LCA) that measures the total primary energy required by a product, service, or system over its entire lifespan. It considers both renewable and non-renewable energy sources, offering insights into the energy efficiency and environmental impact of energy consumption. CED has two primary components, the first being non-renewable energy, which includes energy from finite resources like fossil fuels (coal, natural gas, and oil), nuclear energy, and other non-renewable sources, and the second being renewable energy, which covers energy from sustainable sources such as solar, wind, hydroelectric, geothermal, and biomass energy.
4. **Ozone Depletion Potential (ODP):** The Ozone Depletion Potential (ODP) is a measure in a Life Cycle Assessment (LCA) that assesses how much a substance can damage the ozone layer. The ozone layer is vital for protecting Earth from the sun's harmful ultraviolet (UV) rays. Substances with a high ODP contribute to ozone layer depletion, leading to increased UV radiation reaching Earth, which can harm humans, animals, and ecosystems. Key contributors to ozone depletion include chlorofluorocarbons (CFCs), halons, carbon tetrachloride, methyl chloroform, and some hydrochlorofluorocarbons (HCFCs) and hydro-bromo-fluoro-carbons (HBFCs). These substances emit chlorine and bromine atoms in the stratosphere, which break down ozone molecules.
5. **Particulate Matter Formation (PMF):** Particulate Matter Formation (PMF) is an aspect in a Life Cycle Assessment (LCA) focusing on the potential of emissions to create

particulate matter (PM) in the air. PM comprises small particles or droplets that pose risks to human health as well as to ecosystems and the environment. It varies in size and composition, with PM10 describing particles of 10 µm or less, and PM2.5 describing particles of 2.5 µm or less. Smaller particles are particularly concerning as they can penetrate deep into the respiratory system and even enter the bloodstream. PMF primarily arises from emissions of primary particles, which are emitted directly, and secondary precursors, like sulfur dioxide (SO_2), nitrogen oxides (NOx), ammonia (NH_3), and volatile organic compounds (VOCs), which react in the atmosphere to form particles.

6. **Abiotic Depletion Potential (ADP):** Abiotic Depletion Potential (ADP) is a category in a Life Cycle Assessment (LCA) that examines the potential for depleting non-living (abiotic) resources like minerals and fossil fuels. This depletion is a significant issue due to its impact on the availability of these resources for future generations and the resulting environmental and socio-economic implications. ADP specifically focuses on non-renewable resources, including minerals which encompass metal ores (like iron, copper, and aluminum), industrial minerals (such as limestone and phosphate), and rare earth elements. Also, fossil fuels, with resources like coal, oil, natural gas, and peat, are included in this category.
7. **Photochemical Ozone Creation Potential (POCP):** The Photochemical Ozone Creation Potential (POCP) is a measure used in a Life Cycle Assessment (LCA) to evaluate the likelihood of certain emissions to form ground-level ozone or tropospheric ozone, often referred to as smog. This type of ozone, unlike the protective layer in the upper atmosphere, can negatively impact human health, ecosystems, and crops. Ground-level ozone formation is the result of complex photochemical reactions in the atmosphere, primarily involving volatile organic compounds (VOCs) and nitrogen oxides (NOx). These substances, when released into the air and exposed to sunlight, interact to produce ozone.
8. **Freshwater Ecotoxicity Potential (FETP):** The Freshwater Ecotoxicity Potential (FETP) is an assessment category in a Life Cycle Assessment (LCA) that measures the possible harmful impacts of substances released into freshwater environments. This category evaluates the potential damage to aquatic life in bodies of water like rivers, lakes, and streams, considering both the toxicity and the concentration of the chemicals involved. The substances that contribute to freshwater ecotoxicity vary and include heavy metals, pesticides, industrial chemicals, and pharmaceuticals. These chemicals can negatively affect aquatic organisms by interfering with their biological processes, reproduction, and survival. Such impacts can lead to alterations in the structure and functioning of entire ecosystems.
9. **Human Toxicity Non-Carcinogenic (HTnc):** Human Toxicity Non-Carcinogenic (HTnc) is a category in a Life Cycle Assessment (LCA) focusing on the potential non-carcinogenic adverse health effects on humans from exposure to toxic substances. It addresses a spectrum of health issues, including damage to organs, reproductive and developmental toxicity, neurotoxicity, and endocrine disruption, among others. A wide range of chemicals can contribute to non-carcinogenic human toxicity, including heavy metals, solvents, pesticides, industrial chemicals, and air pollutants. These substances can be absorbed into the human body via inhalation, ingestion, or skin contact, and the resulting health impacts vary based on the amount (or dose), duration, and method of exposure.
10. **Human Toxicity carcinogenic (HTc):** Human Toxicity Carcinogenic (HTc) is a crucial impact category in a Life Cycle Assessment (LCA) that assesses the potential health risks associated with exposure to carcinogenic substances. These substances may cause cancer in living tissues, representing a significant health hazard. The assessment looks at different pathways of exposure, such as inhalation, ingestion, and skin contact. Various substances are identified as potential contributors to carcinogenic human toxicity, including polycyclic aromatic hydrocarbons (PAHs), volatile organic

compounds (VOCs) like benzene, heavy metals (for example, arsenic, cadmium, and chromium), asbestos, formaldehyde, dioxins, and furans, as well as some pesticides and herbicides known to have carcinogenic effects.

11. **Eutrophication Potential for terrestrial (EPt):** The Eutrophication Potential for Terrestrial Ecosystems (EPt) in a Life Cycle Assessment (LCA) evaluates the environmental impacts of excessive nutrient enrichment in land ecosystems. This phenomenon, primarily caused by nitrogen and phosphorus compounds, leads to changes in soil chemistry, alterations in plant communities, and habitat degradation. The major contributors to this issue include nitrogen compounds (like ammonia, nitrogen oxides, and nitrates) and phosphorus compounds (such as phosphates), originating from agriculture, industry, transport, and waste management. This LCA category helps in assessing and mitigating the impacts of nutrient overloading on terrestrial environments.
12. **Eutrophication Potential for Marine Ecosystems (EPm):** The Eutrophication Potential for Marine Ecosystems (EPm) in a Life Cycle Assessment (LCA) focuses on assessing the environmental impacts of excessive nutrient enrichment in oceanic habitats. This enrichment, primarily from nitrogen and phosphorus compounds, can cause issues like harmful algal blooms, oxygen depletion (hypoxia), biodiversity loss, and changes in marine habitats. Nitrogen compounds (such as nitrate and ammonia) and phosphorus compounds (like phosphate) are the main contributors. These nutrients typically come from agricultural runoff, wastewater discharge, industrial emissions, and atmospheric deposition. This LCA category helps in understanding and managing the ecological impacts on marine environments due to nutrient overloading.
13. **Water Depletion**: Water Depletion is a significant impact category in a Life Cycle Assessment (LCA) that aims to evaluate the potential environmental impacts associated with the depletion of freshwater resources. Water Depletion considers both the quantity and quality aspects of water consumption and contamination, assessing the stress placed on water resources and the consequent ecological, societal, and economic implications.
14. **Land Use and Land Change:** Land Use and Land Use Change are crucial categories in a Life Cycle Assessment (LCA) for evaluating the environmental impact of using and altering land for human activities. "Land Use" examines the impact of using land for agriculture, forestry, urban, or industrial purposes, focusing on the duration and intensity of use and its effects on biodiversity, soil, and ecosystem services. "Land Use Change" deals with the transformation of land from one type to another, such as from forests to farmland or from grasslands to urban areas, and its implications on land cover, habitat loss, albedo changes, and carbon and water cycles.
15. **Biodiversity Loss**: Biodiversity Loss is a key impact category in a Life Cycle Assessment (LCA) that investigates the potential adverse effects of human activities on the variety of life on Earth, including the different species of plants, animals, and microorganisms, the genetic differences within these species, and the ecosystems they form.
16. **Noise Pollution**: Noise Pollution is an essential impact category in a Life Cycle Assessment (LCA) that focuses on evaluating the environmental and human health impacts associated with unwanted or harmful sound levels produced during various life cycle stages of products, services, or systems. It is a significant concern due to its potential effects on human health, well-being, wildlife, and the overall quality of the environment.
17. **Soil Quality Degradation:** Soil Quality Degradation is a crucial impact category in a Life Cycle Assessment (LCA) that addresses the decline in the health and functionality of soil as a result of human activities. Soil quality is integral to ecosystem services as good soil quality supports plant growth, regulates water flow, cycles nutrients, and hosts a vast array of biodiversity.
18. **Thermal Pollution**: Thermal Pollution is an important impact category in a Life Cycle Assessment (LCA) that assesses the effects of abnormal changes in the environmental

temperature due to human activities. It typically occurs when industries or power plants discharge heated water or air into the environment, affecting water quality and ecosystems, particularly aquatic life.

19. **Groundwater Contamination:** Groundwater Contamination is a critical impact category in a Life Cycle Assessment (LCA) that evaluates the extent and implications of pollutants entering groundwater resources due to human activities. Groundwater is a vital source of drinking water and irrigation, and its contamination can have severe repercussions on human health, ecosystems, and water availability.

Inclusions and Exclusions of Environmental Impact Factors

In this study, the SimaPro software V7 was first used with its ReCiPe Endpoint (H) V1.06/World ReCiPe H/H method for an environmental impact assessment. This approach allowed for the analysis of 12 out of 19 potential impact categories. Due to software constraints, it was not possible to include the remaining seven critical factors: Global Warming Potential, Cumulative Energy Demand, Water Depletion, Noise Pollution, Soil Quality Degradation, Thermal Pollution, and Groundwater Contamination. In order to provide a full consideration of the LCA effects, a separate discussion about these factors is later provided. The subsequent use of the GREET software V2 provided a broader perspective, offering detailed data on emissions and energy consumption at various stages of the recycling process. This facilitated a more comprehensive understanding of the Global Warming Potential, Cumulative Energy Demand, and Water Depletion. Nevertheless, four elements—Noise Pollution, Soil Quality Degradation, Thermal Pollution, and Groundwater Contamination—remained elusive due to insufficient available data, making their quantification challenging. Table 1 categorizes all 19 environmental factors, indicating those analyzed and those beyond the study's scope, for full consideration.

Table 1. Inclusion and exclusion of environmental factors.

Environmental Factors	Included in SimaPro	Included in GREET	Excluded from Both
Global Warming Potential		X	
Acidification Potential	X		
Cumulative Energy Demand		X	
Ozone Depletion Potential	X		
Particulate Matter Formation	X		
Abiotic Depletion Potential	X		
Photochemical Ozone Depletion Potential	X		
Human Toxicity Non-Carcinogenic	X		
Human Toxicity Carcinogenic	X		
Eutrophication Potential for Terrestrial Ecosystems	X		
Eutrophication Potential for Marine Ecosystems	X		
Water Depletion		X	
Land Use and Land Change	X		
Biodiversity Loss	X		
Noise Pollution			X
Soil Quality Degradation			X
Thermal Pollution			X
Groundwater Contamination			X

4. LCA Simulation

4.1. LCA Software Considered

Two software platforms, SimaPro and GREET, were selected for this Life Cycle Assessment (LCA) analysis. SimaPro stands out for its detailed LCA capabilities, including a rich database such as Eco-invent and support for various impact assessment methods, including Eco-Indicator, EDIP, EPD, ReCiPe, and CML. This set of capabilities positions it alongside other leading LCA software like GaBi Pro and OpenLCA, enabling the direct computation of environmental impacts to evaluate the effects on the environment and human health. Its industry-leading status is bolstered by its robust database.

GREET, in contrast, is a freely available software tool developed by the Argonne National Laboratory with support from the U.S. Environmental Protection Agency. It distinguishes itself with customizable LCA features, particularly in presenting results as actual emissions for each life cycle stage, and it facilitates a comparative, detailed analysis of different recycling methods in this application. When used together, SimaPro and GREET provide a comprehensive view of the entire LCA process, enhancing the accuracy of assessments regarding the impact of recycling EV batteries on the environment and human health.

4.2. SimaPro

The original SimaPro software was released in 1990. It is a leading software tool for Life Cycle Assessments (LCAs) and is widely used in industry, consulting, and academia. It offers comprehensive LCA capabilities, extensive environmental impact databases like Eco-invent, and supports multiple impact assessment methods such as ReCiPe and CML. Its flexibility allows for customized studies and scenario analyses to be conducted, and it also facilitates collaboration and detailed reporting. Additionally, SimaPro can be integrated with other tools for advanced analyses, making it a valuable asset for evaluating the environmental impact of products and services, leading to potential process modifications for impact reduction.

To model an EV battery's life cycle using SimaPro, the nickel–cobalt–manganese (NCM) chemistry was chosen as a representative of the current supply leader. The assessment used the battery composition shown in Table 2.

Table 2. NCM battery composition.

Mineral	Cell Part	Average Content in kg	Content % of Total
Graphite	Anode	52	28.1%
Aluminum	Cathode, Case, Current Collectors	35	18.9%
Nickel	Cathode	29	15.7%
Copper	Current Collectors	20	10.8%
Steel	Case	20	10.8%
Manganese	Cathode	10	5.4%
Cobalt	Cathode	8	4.3%
Lithium	Cathode	6	3.2%
Iron	Cathode	5	2.7%
Total		**185 kg**	**100%**

An Assembly, sourced from SimaPro's Eco-invent library, was created within the SimaPro software to simulate the battery. Because the library did not allow for an exact duplication of materials, as described in Table 1, material equivalents for some components were selected and are shown in Table 3.

Table 3. SimaPro battery assembly.

EV Battery Materials	Equivalent
Anode, lithium-ion battery, graphite, at plant	Graphite
Nickel, 99.5%, at plant	Nickel
Cathode, copper, primary copper production	Copper
Steel, converter, chromium steel 18/8, at plant	Steel
Cathode, lithium-ion battery, lithium manganese oxide, at plant	Lithium and Manganese
Cobalt, at plant	Cobalt
Aluminum alloy, $AlMg_3$, at plant	Aluminum
Iron-nickel-chromium alloy, at plant	Iron

SimaPro enables the integration of production process impacts into the final product analysis. In the case of this assembly, focus was exclusively placed on the battery manufacturing process. With New Jersey used as the benchmark location for this study and considering that natural gas constitutes the primary energy source in the stat (accounting for 46% of energy production [19]), the assumptions were made based on the premise that all of the energy utilized in the battery production facility is derived from natural gas. According to Kim et al. [20], energy use for cell production and battery pack manufacturing amounts to 1500 MJ/kWh, which equates to 28 kg/kWh of natural gas. This amount was divided equally between two processes in SimaPro, with the first being energy for factory operation and the second being energy for machine operation, with 14 kg/kWh of natural gas each. The difference between the energy spent on machine operations and factory operations lies in their scope. Energy used in machine operations is specific to the power consumed by the production equipment and machinery during their active use. In contrast, energy spent on factory operations is more comprehensive, covering all energy usage within the factory. This broader scope includes essential facilities like lighting, heating, cooling, and ventilation, encompassing the overall operational energy requirements of the factory environment. For the disposal scenario, because of the limitations inherent in the available library data, only incineration was considered as a recycling option.

As mentioned previously, for the impact assessment analysis of this assembly, the ReCiPe Endpoint (H) V1.06/World ReCiPe H/H method was considered, which is SimaPro's most comprehensively used method. The impact categories that are utilized in this methodology target 12 of the 19 impact categories of interest that were discussed earlier, with Global Warming Potential, Cumulative Energy Demand, Water Depletion, Noise Pollution, Soil Quality Degradation, Thermal Pollution, and Groundwater Contamination not being accounted for in this particular SimaPro approach.

The SimaPro analysis showed that nickel, copper, and graphite are the primary contributors to environmental impacts during battery production. This important environmental footprint aligns with the usage of these raw materials in battery manufacturing, where graphite is the most used material, followed by nickel and then copper. The impact dominance of these materials is further explained by the environmentally detrimental methods of open-pit and strip mining that are commonly used for their extraction.

In the normalized results for the battery, Human Toxicity emerged as the most significant environmental impact, followed by Particulate Matter Formation, Fossil Fuel Depletion, and Climate Change Human Health. These impacts are largely due to the environmentally negative practice of open-pit mining that is employed in extracting these minerals. Open-pit mining in general leads to deforestation, habitat destruction, and soil erosion, and it leaves excavated sites barren, often without significant efforts to restore the lost vegetation. Additionally, the blasting frequently used during mining generates fine dust that also poses serious health risks. For example, South Africa's mining sector reaches up to ten times the emergency threshold defined by the World Health Organization, with 2500–3000 cases of tuberculosis per 100,000 individuals, due in part to high levels of particulates in mining operations [21]. As a result of the open-air open-pit mining operations, fine dust, a form of particulate matter, cannot be effectively contained and destroyed.

An examination of the environmental impacts of incineration relative to battery manufacturing revealed that the impacts of manufacturing significantly outweigh those of incineration. The presence of baghouses in incineration plants effectively mitigates particulate matter emissions. These systems work by pulling in air that is laden with particles, filtering these particles out, and then either releasing clean air back into the environment or reusing it within the plant. However, the primary environmental concern with incineration lies in its contribution to climate change, primarily through the emission of greenhouse gases, which remains a challenge despite particle filtration.

Lastly, it is crucial to acknowledge a key limitation of this study: the reliance on SimaPro's libraries that were last updated in 2010. This constraint not only limits the functionalities available in the software but also affects the currency and temporal relevance of the information generated. So, while trends can be identified, the precise delineation of impacts cannot be obtained using this database.

4.3. Using GREET

GREET (Greenhouse Gases, Regulated Emissions, and Energy Use in Transportation) is another LCA software tool, which was developed by the Argonne National Laboratory of the U.S. Department of Energy. First released in the mid-1990s, GREET was designed to evaluate the energy and environmental impacts of various vehicle technologies and transportation fuels over their entire life cycles. The generated model provides a comprehensive analysis of various factors, including energy consumption, greenhouse gas emissions, and air pollution. GREET is widely used for research and policy analysis, helping to inform decisions in transportation, energy policy, and environmental impact assessments. Its regular updates and enhancements have made it a pivotal tool in understanding the complex interactions between transportation technology and environmental outcomes.

In this research, two electric vehicles were considered: the Tesla Model 3, with a battery weight of 480.8 kg, and the Nissan Leaf, with a battery weight of 303 kg. Utilizing the GREET software V2, these vehicles were modeled based on the EV300—Electricity (Type 1 Li-Ion/NMC111 Conventional Material) vehicle template from GREET's library. A key feature of GREET is its ability to customize the electricity mix used for charging the vehicles. For this analysis, the electricity composition specific to New Jersey was applied, as depicted in Figure 1. This composition is based on the data provided by the EIA [19].

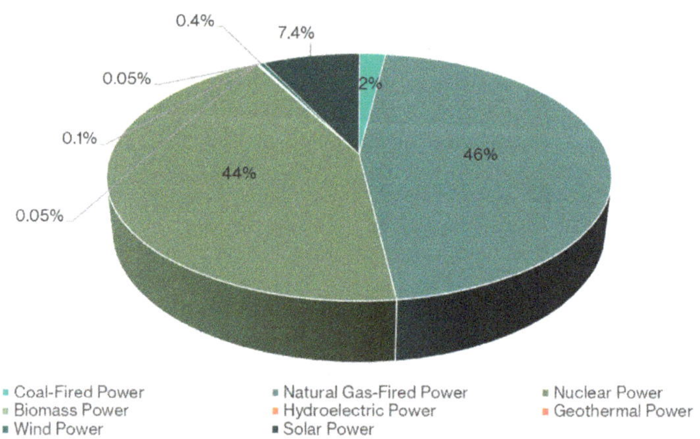

Figure 1. Energy distribution in New Jersey.

When analyzing the results from GREET, only variables of interest for this situation are, namely disposal and recycling and battery assembly and manufacturing. These factors will be used here to illustrate the differences in impacts between producing a new battery from raw materials and those generated by producing a battery using recovered materials obtained by battery recycling. The results from GREET are displayed across twenty-three columns, each representing different aspects of environmental impact and energy use related to transportation. A brief description of each column is provided as follows:

1. **Total Energy (J/mi):** This measures the total energy consumed per mile, encompassing all energy sources.
2. **Fossil Fuel (J/mi):** This indicates the amount of energy derived from fossil fuels that is used per mile.
3. **Coal Fuel (J/mi):** This shows the energy from coal used per mile.
4. **Natural Gas Fuel (J/mi):** This represents the energy obtained from natural gas used per mile.
5. **Petroleum Fuel (J/mi):** This denotes the energy from petroleum products used per mile.
6. **Renewable (J/mi):** This represents the amount of renewable energy used per mile.
7. **Biomass (J/mi):** This represents energy derived from biomass used per mile.
8. **Nuclear (J/mi):** This represents energy from nuclear sources used per mile.
9. **Non-Fossil Fuel (J/mi):** This represents the energy from non-fossil sources used per mile.
10. **VOC (kg/mi):** This represents the emissions of volatile organic compounds per mile.
11. **CO (kg/mi):** This represents carbon monoxide emissions per mile.
12. **NOx (kg/mi):** This represents nitrogen oxide emissions per mile.
13. **PM10 (kg/mi):** This represents particulate matter (10 μm or less) emissions per mile.
14. **PM2.5 (kg/mi):** This represents fine particulate matter (2.5 μm or less) emissions per mile.
15. **SOx (kg/mi):** This represents sulfur oxide emissions per mile.
16. **CH_4 (kg/mi):** This represents methane emissions per mile.
17. **CO_2 (kg/mi):** This represents carbon dioxide emissions per mile.
18. **N_2O (kg/mi):** This represents nitrous oxide emissions per mile.
19. **BC (kg/mi):** This represents black carbon emissions per mile.
20. **POC (kg/mi):** This represents primary organic carbon emissions per mile. (Basically, these are combustible carbon compounds that can be filtered from emissions.)
21. **CO_2_Biogenic (kg/mi):** This represents biogenic carbon dioxide emissions per mile. (Basically, this is CO_2 derived from biological sources other than fossil fuels.)
22. **GHG-100 (kg/mi):** This represents greenhouse gas emissions with a 100-year global warming potential per mile.
23. **GHG-20 (kg/mi):** This represents greenhouse gas emissions with a 20-year global warming potential per mile.

The results yielded from the GREET software V2 for the Tesla Model 3 are shown in Table 4. The data presented in both Table 4 and Figure 2 clearly indicate that manufacturing new (virgin) batteries demands significantly more energy compared to using recycled material in existing batteries. Here, a comparison is made between the production of a virgin battery with the sum of the impacts of the recycling process and the manufacture of a new battery using recovered materials. Specifically, the energy consumption for manufacturing virgin batteries is eight times higher than that based on recycling, emphasizing a substantial disparity between the two processes. This finding highlights the critical importance of investing in recycling facilities, which can lead to considerable energy savings with resulting reductions in environmental impacts. Considering the energy sources in New Jersey, where energy consumption (75.1 TWhrs) surpasses production (65.3 TWhrs), as reported by the U.S. Department of Energy in 2016 [22], and with the prospect of rising energy prices as a result of meeting escalating electricity demands, it becomes increasingly essential to explore innovative approaches for reducing energy consumption.

Table 4. Tesla Model 3—energy emissions.

Name	Disposal and Recycling	Virgin Battery
Total Energy (J/mi)	53,888	434,050
Fossil Fuel (J/mi)	44,512	379,658
Coal Fuel (J/mi)	1041	65,383
Natural Gas Fuel (J/mi)	43,417	260,341
Petroleum Fuel (J/mi)	54	53,934
Renewable (J/mi)	1470	23,001
Biomass (J/mi)	41	1293
Nuclear (J/mi)	7906	31,390
Non-Fossil Fuel (J/mi)	9376	54,391

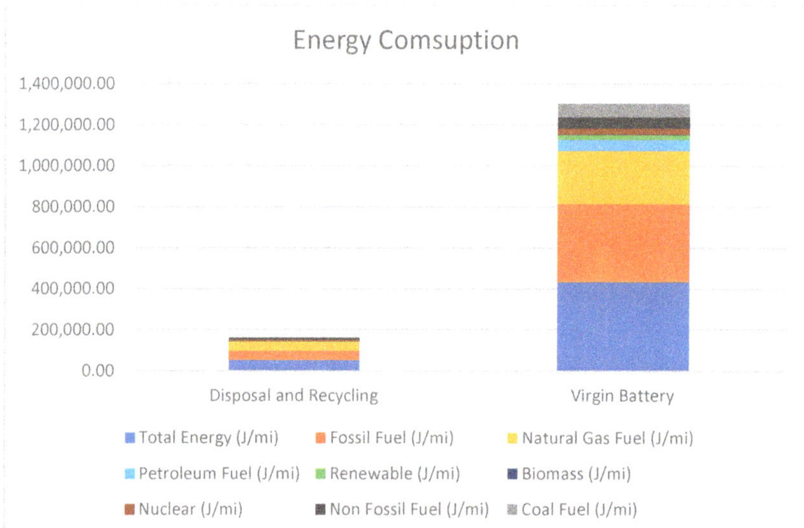

Figure 2. Tesla Model 3 Energy Consumption—recycling vs. virgin battery.

When looking at the energy consumption for manufacturing a battery using virgin raw material versus the energy consumption when using recycled Nissan Leaf batteries, it can be seen that the data paint a very similar story to that seen with the Tesla Model 3 (see Table 4 and Figure 2), with the total energy needed to produce a new battery also being also eight times greater than that needed for raw materials obtained by battery recycling, as shown in Table 5 and Figure 3.

Regarding the environmental emissions of both cars, starting with the Tesla Model 3, as can be seen in Table 6 and Figure 4 the environmental emissions from virgin battery manufacturing far outweigh the total emissions produced by using raw material obtained from battery recycling. One important aspect to note is that CO_2, GHG-20, and GHG-100 are greater than all other impact categories listed, as can be clearly seen in Figure 4. Moreover, for all three categories, virgin battery manufacturing produces about 10 times more emissions than recycling, further highlighting the benefits that the recycling process has.

Table 5. Nissan Leaf—energy emissions.

Name	Disposal and Recycling	Virgin Battery
Total Energy (J/mi)	34,037.09	275,611.2
Fossil Fuel (J/mi)	28,114.89	241,131.1
Coal Fuel (J/mi)	657.43	413,38.32
Natural Gas Fuel (J/mi)	27,423.05	165,701.8
Petroleum Fuel (J/mi)	34.41	34,091
Renewable (J/mi)	928.48	14,554.24
Biomass (J/mi)	26.21	817.8613
Nuclear (J/mi)	4993.72	19,925.82
Non-Fossil Fuel (J/mi)	5922.21	34,480.06

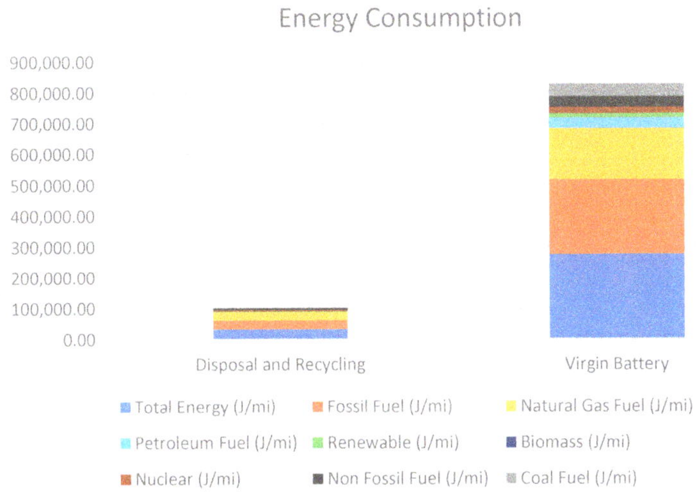

Figure 3. Nissan Leaf Energy Consumption —recycling vs. virgin battery.

Table 6. Tesla Model 3—environmental impacts.

Name	Disposal and Recycling	Virgin Battery
VOC (kg/mi)	6.74×10^{-6}	7.10×10^{-6}
CO (kg/mi)	1.91×10^{-6}	4.70×10^{-5}
NOx (kg/mi)	2.57×10^{-6}	3.11×10^{-5}
PM10 (kg/mi)	3.87×10^{-7}	1.39×10^{-5}
PM2.5 (kg/mi)	2.59×10^{-7}	4.48×10^{-6}
SOx (kg/mi)	5.36×10^{-7}	7.23×10^{-5}
CH_4 (kg/mi)	7.88×10^{-8}	1.32×10^{-4}
CO_2 (kg/mi)	0.002533	0.0242
N_2O (kg/mi)	6.39×10^{-6}	7.06×10^{-7}
BC (kg/mi)	1.77×10^{-8}	2.55×10^{-7}

Table 6. *Cont.*

Name	Disposal and Recycling	Virgin Battery
POC (kg/mi)	6.77×10^{-8}	6.00×10^{-7}
CO_2_Biogenic (kg/mi)	-3.81×10^{-6}	-1.17×10^{-4}
GHG-100 (kg/mi)	0.0028	0.0284
GHG-20 (kg/mi)	0.003267	0.0353

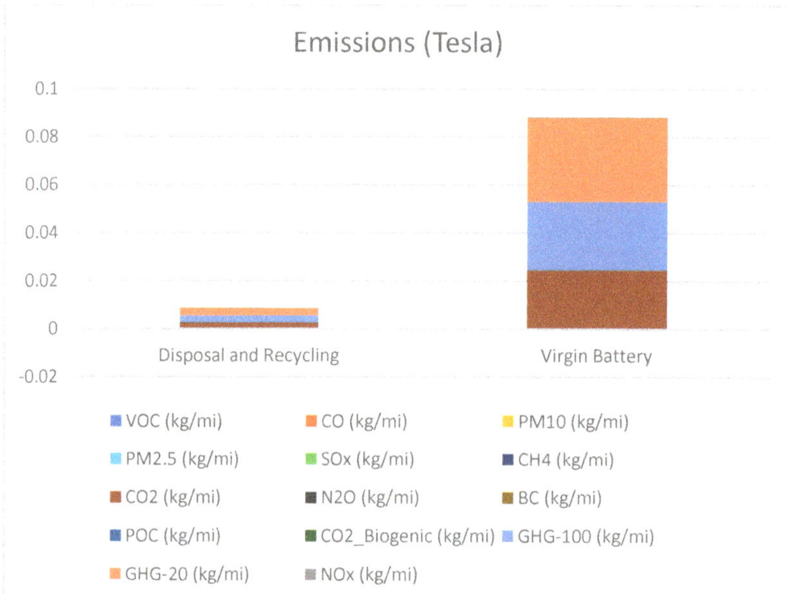

Figure 4. Tesla Model 3 Emissions—recycling vs. virgin battery.

The results obtained for the Nissan Leaf model are very similar to those obtained for the Tesla Model 3, as can be seen in Table 7 and Figure 5.

A notable limitation of the methodology used in this study is related to how GREET handles the computation of disposal and recycling. The software automatically depends on its internal database to calculate ADR (Assembly, Disposal, and Recycling). To simplify the conducted analysis, this research assumed an equal distribution across these three components to isolate disposal and recycling. However, this approach restricts the ability to modify or specifically control certain parameters within these calculations. This constraint becomes particularly challenging when attempting to incorporate and analyze specific recycling processes, such as hydrometallurgy or pyrometallurgy, within GREET. To address this limitation, it was assumed that the GREET database accurately represents vehicle materials for recycling and disposal. This simplification helped to navigate the software's constraints, enabling this research to proceed by focusing on an equal distribution across ADR components. However, it is clear that a different approach would be required to assess differences among the specific processes that might be used for EV battery recycling.

Table 7. Nissan Leaf—environmental impacts.

Name	Disposal and Recycling	Virgin Battery
VOC (kg/mi)	6.04×10^{-7}	4.50×10^{-6}
CO (kg/mi)	2.46×10^{-6}	2.98×10^{-5}
NOx (kg/mi)	3.33×10^{-6}	1.98×10^{-5}
PM10 (kg/mi)	1.81×10^{-7}	8.81×10^{-6}
PM2.5 (kg/mi)	1.76×10^{-7}	2.84×10^{-6}
SOx (kg/mi)	5.76×10^{-7}	4.57×10^{-5}
CH_4 (kg/mi)	1.00×10^{-5}	8.36×10^{-5}
CO_2 (kg/mi)	0.0031	0.0154
N_2O (kg/mi)	9.06×10^{-8}	4.48×10^{-7}
BC (kg/mi)	2.66×10^{-8}	1.62×10^{-7}
POC (kg/mi)	7.88×10^{-8}	3.81×10^{-7}
CO_2_Biogenic (kg/mi)	-1.90×10^{-6}	-7.41×10^{-5}
GHG-100 (kg/mi)	0.0034	0.018
GHG-20 (kg/mi)	0.0039	0.0224

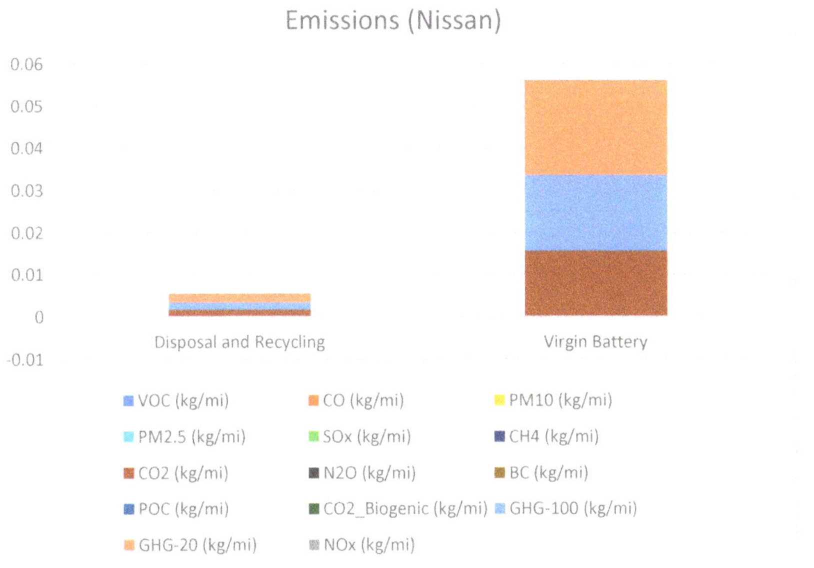

Figure 5. Nissan Leaf Emissions—recycling vs. virgin battery.

4.4. *Pros and Cons of LCA Software*

SimaPro V7 offered several advantages, primarily its automated generation of environmental impacts, which enables users to quantify various environmental aspects using a diverse range of methods and libraries. These methods include the ReCiPe method for global assessment and the CML 2 method tailored to the European market. This versatility allows researchers to focus on specific environmental aspects, such as Global Warming Potential, Acidification Potential, and the Eutrophication of Freshwater, among others. SimaPro also generates informative graphs and tree networks to visualize the results and their breakdown, facilitating a deeper understanding of the environmental impacts. Furthermore, users can normalize the results to identify dominant environmental factors. SimaPro's capacity to create and compare different assemblies and LCA systems is another valuable feature. Table 8 shows a comparison of SimaPro with GREET.

Table 8. Comparison between SimaPro and GREET.

Feature/Aspect	SimaPro	GREET
Environmental Impact Generation	Automated generation of environmental impacts using diverse methods and libraries	Offers actual emission data including total energy expenditure and specific emissions (e.g., CO_2, CH_4, and VOCs)
Methodologies	Includes ReCiPe and CML 2 methods, which are suitable for global and European market assessments	Users can tailor entire LCA model, including energy sources, manufacturing processes, and recycling methods
Focus Areas	Focuses on Global Warming Potential, Acidification Potential, Eutrophication, etc.	Customizable focus on various aspects of lifecycle, including raw material extraction and disposal
Visualization Tools	Provides graphs and tree networks for visualizing and understanding environmental impacts	Lacks advanced visualization tools; relies on raw data presentation
Result Normalization	Enables normalization of results to identify dominant environmental factors	Does not inherently provide normalization of results for environmental impact assessment
Database Timeliness	Limited by timeliness of its database, with no user access to update it with latest data	Allows for custom input, adapting to latest data and techniques in rapidly evolving fields
Customization	Limited customization in creating and comparing different assemblies and LCA systems	Exceptional customizability in energy mixes, manufacturing processes, transportation, and disposal techniques
Quantification of Impacts	Effective at quantifying environmental impacts relevant to human health and ecosystems	Requires external methods (like APEEP Model) to quantify impacts on human health and environment
Suitability	Versatile for researchers focusing on specific environmental aspects	Suitable for users needing high customizability and detailed emission data

However, SimaPro has some major drawbacks due to its heavy reliance on the current version of its database because users have no access to updates or later versions, and it also provides a comparison or offers ratios of different environmental categories and not actual data. These limitations become particularly problematic in rapidly evolving fields, such as recycling, where new data and techniques regularly emerge.

In contrast, GREET stands out due to its exceptional customizability. Users can specify energy sources, create unique energy mixes, simulate various manufacturing processes, define transportation modes and methods, and customize recycling and disposal techniques. GREET also provides an extensive library of pre-existing products that serve as starting points for customization. The entire LCA model, from raw material extraction to manufacturing and recycling, can be tailored to specific needs.

Nonetheless, GREET's main disadvantage lies in how it presents results. Unlike SimaPro, GREET offers actual emission data for the entire process, including total energy expenditure and emissions like carbon dioxide (CO_2), methane (CH_4), and volatile organic compounds (VOCs), among others. This format is less useful when quantifying environmental impacts, especially those related to human health. Users are compelled to rely on external methods like the APEEP (Air Pollution Emission Experiments and Policy) Model to transform these emissions into quantifiable environmental impacts [23].

4.5. Qualitative Assessment

Software applications, such as GREET and SimaPro, are well-suited for analyzing environmental impacts in areas with an abundance of quantitative data, including data on Acidification Potential and Ozone Depletion Potential. These tools rely on extensive and comprehensive databases to model and assess the environmental footprints of various processes and products. However, for factors such as Noise Pollution, Soil Quality Degradation, Thermal Pollution, and Groundwater Contamination, the variability and scarcity of data across different projects pose significant challenges. In such cases, the standardized databases and algorithms used by these software applications may not provide accurate or relevant insights, rendering their application less effective. Despite this, the qualitative assessments these software approaches can provide yield insights into the environmental impacts of the products and services being studied. Consequently, the subsequent section will undertake a qualitative analysis to evaluate these factors' environmental effects.

While there are no specific data when it comes to noise pollution in the context of EV battery manufacturing and recycling, results can be inferred from information about car manufacturing. Comparing the noise pollution from the production of new cars to the recycling of old ones requires evaluating the intensity and duration of noise generated by each process. The manufacturing process of new cars involves heavy machinery for metal pressing, welding, and assembly, which can generate noise levels up to 85–90 decibels (dB) or more depending on the specific operations and machinery used [24]. This process is continuous, leading to sustained noise pollution. On the other hand, the recycling of cars involves dismantling, shredding, and reprocessing materials, with noise levels potentially reaching up to 123 dB during the most intensive operations like metal shredding [25]. However, these activities may not be as continuous as car manufacturing, potentially resulting in less sustained, but more intense, periods of noise pollution. The impact of noise also heavily depends on the proximity of these activities to residential areas and the presence and effectiveness of noise mitigation measures. While both processes generate significant noise, manufacturing might contribute to more consistent noise pollution due to its continuous nature, whereas recycling operations can have peak noise levels that are higher during specific activities but might not be as constant.

The impact of producing new LIBs on soil quality versus the impact of making them using recycled materials involves different aspects of environmental interaction. LIB manufacturing using virgin supplies entails extensive resource extraction, including mining for metals and minerals like lithium and cobalt, which can significantly degrade soil quality. These activities lead to soil erosion, heavy metal contamination, and changes in soil composition due to the disposal of industrial waste [26]. On the other hand, LIB recycling can also affect soil quality, primarily through the potential leakage of hazardous substances such as PFAS, lead, mercury, and cadmium during the dismantling process [27]. However, recycling aims to reduce waste and reuse materials, which can mitigate some soil degradation by decreasing the demand for new raw materials and minimizing the footprint of waste disposal [28]. While both processes have the potential to impact soil negatively, the scale and nature of their impacts differ. Manufacturing new LIBs has a broader environmental footprint that includes the degradation of soil quality at resource extraction sites and around manufacturing plants [29]. In contrast, the impact of recycling is more localized and can be mitigated through proper waste management practices and environmental safeguards. Overall, EV battery recycling, when conducted responsibly, tends to have a less detrimental impact on soil quality compared to the extensive soil degradation associated with the resource extraction and waste production from manufacturing new batteries.

When comparing the thermal pollution associated with lithium-ion battery (LIB) manufacturing versus LIB recycling, it is important to consider the energy-intensive processes involved in both. Manufacturing LIBs from new materials is significantly more energy and heat-intensive compared to producing them from recycled materials, leading to higher thermal pollution. This is because new battery production involves energy-consuming stages, such as material extraction and processing, while recycling reduces the need for these processes by reusing materials. Although recycling also requires energy, particularly for dismantling and chemical treatment, it generally consumes less energy than manufacturing, resulting in lower thermal pollution [30,31]. Consequently, while both manufacturing and recycling LIBs contribute to thermal pollution, the environmental impact of new battery production from virgin raw materials is considerably greater than that using raw materials obtained from battery recycling, highlighting the benefits of recycling in mitigating thermal pollution.

Groundwater contamination risks differ markedly between the production and recycling of lithium-ion batteries (LIBs). During the production phase, the extraction and processing of raw materials such as lithium, cobalt, and nickel can lead to the release of toxic chemicals into the environment, potentially contaminating groundwater sources. These activities often involve the use of hazardous chemicals for metal extraction and processing, which, if not properly managed, can seep into soil and groundwater [32]. On the recycling side, while there is a potential for groundwater contamination through the improper handling and disposal of battery components, advanced recycling processes aim to minimize this risk by safely extracting valuable materials and treating waste products. Recycling facilities are increasingly adopting measures to prevent the leakage of hazardous substances, thus mitigating the risk of groundwater contamination. However, the effectiveness of these measures depends on the particular recycling technologies used and the regulatory frameworks in place. The U.S. Environmental Protection Agency (EPA) has provided guidance on how to handle hazardous waste from lithium-ion batteries under the Resource Conservation and Recovery Act (RCRA) [33]. Overall, while both the production and recycling of LIBs pose risks to groundwater quality, the managed environment of recycling processes, when conducted according to best practices, tends to present a lower risk of contamination compared to the extensive environmental impact associated with raw material extraction and processing as a part of battery production.

Therefore, the qualitative assessment of these factors is aligned with what has been found quantitatively. Recycling decreases the environmental costs associated with the production of virgin batteries from raw material extraction to manufacturing. However, a qualitative assessment adds another layer to this analysis, highlighting how manufacturing can disrupt and reduce the quality of life of citizens, with more pollution that they are forced to deal with, as well as more polluted water and soil, which can have detrimental effects on the health of these citizens, especially children. Hence, this qualitative analysis further supports the use of recycling.

5. Life Cycle Cost Analysis

A Life Cycle Cost Analysis (LCCA) plays a pivotal role in the realm of electric vehicle (EV) battery recycling. By meticulously evaluating the costs associated with the entire life cycle of battery recycling, an LCCA provides invaluable insights into the economic feasibility and sustainability of the recycling process. The ultimate objective of employing an LCCA in this context is to accurately determine the cost of recycling EV batteries. This, in turn, is crucial for calculating the expense involved in manufacturing new batteries from recycled materials. Such an analysis is essential not only for understanding the economic implications of recycling, but also for promoting more sustainable and cost-effective developments in the rapidly evolving EV industry. It helps in making informed decisions that balance environmental benefits with financial viability, thereby contributing to a more sustainable future in the automotive sector.

In this research, the cost of recycling electric vehicle batteries was explored and compared to the cost of new batteries. The price of a new Tesla Model 3 battery is currently USD 15,800, and a Nissan Leaf battery is priced at USD 6500 [34]. To estimate the recycling costs, two methods were employed. The first method, proposed by the Argonne National Laboratory (the developers of the GREET software V2) in 2000 [35], initially set the recycling cost at USD 10 per kilogram of battery. This cost has since decreased to USD 5 per kilogram. This reduction in cost can be attributed to advancements in recycling technologies, increased efficiency in the recycling process, and economies of scale with the growth in the volume of batteries to be recycled. As illustrated in Figure 6, recycled raw material proves to be substantially more economical than purchasing virgin raw materials for the production of new batteries. Specifically, for a Tesla battery, the acquisition of recycled materials is six times less expensive than obtaining virgin materials, and for a Nissan Leaf battery, the differential is four times cheaper.

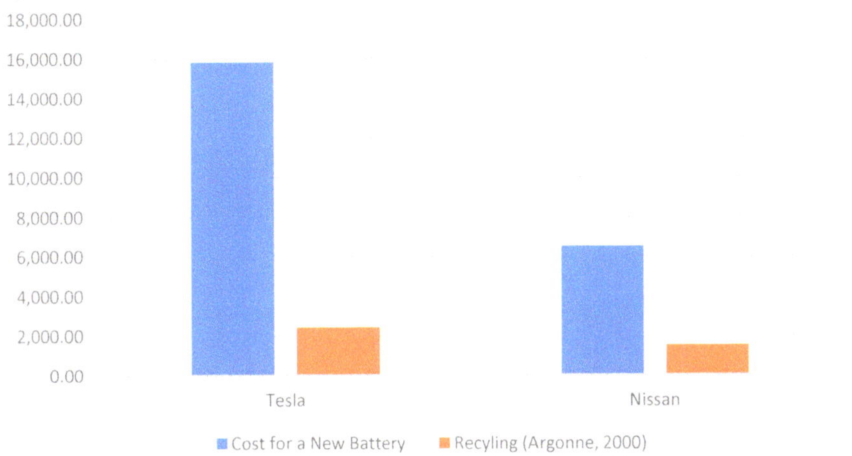

Figure 6. Creating new batteries vs. recycling.

The second approach to estimating battery recycling costs is derived from a 2019 report by the Argonne National Laboratory [36]. According to this report, the current total cost for recycling lithium batteries is approximately USD 26 per kWh. This includes a recycling fee of USD 10 per kWh, a charge applied by recyclers for the collection, processing, and recovery of materials from spent batteries. Figure 7 contrasts the price of a new battery (blue bar) to that of a recycled battery, which was calculated using the formula provided above, developed by the Argonne Laboratory (orange bar). Based on this pricing, the cost to produce a Tesla Model 3 battery is calculated to be USD 1692, while that for a Nissan Leaf battery is USD 1066. These figures represent a significant reduction compared to the cost of manufacturing batteries using virgin raw materials. This further suggests that recycling should be recognized as a more cost-effective option when comparing the results to what is presented in Figure 6, which is likely due to advancements in recycling technology and processes becoming more streamlined and efficient over time, coupled with increased investments in the field. This trend reinforces the potential for even lower manufacturing costs in the future when using recycled materials.

Patrick Curran, a GLG Network Member and the CEO of Lithium Recycling Systems, has provided insightful data on the economics of battery recycling [37]. He noted that processing one metric ton of incoming batteries costs around USD 90. From this process, the black mass obtained—a combination of nickel, manganese, and cobalt oxides with carbon—can be sold for approximately USD 300 or more. Additionally, the metallic components, mainly copper and aluminum found in the batteries, can fetch around USD 500. These figures point to recycling as not only a profitable venture for investors, but also as a cost-effective component of battery production. Recycling batteries to create new batteries is substantially less expensive than using virgin materials, as demonstrated in Figure 8. This highlights the economic advantage and potential savings achievable through the use of recycled materials in the battery manufacturing industry. The analysis performed here is rudimentary, and the economics are likely slightly more complicated for the following reasons. Assuming the same level of purity and regardless of the source, the materials—lithium, cobalt, copper, etc.—will be sold at market value. So, if the organization that is responsible for collecting spent batteries and processing them to recover the materials also makes the replacement battery, then they will have larger cost savings. Otherwise, if they sell the materials to a manufacturer, they will make a profit, but the price of the manufactured battery will be comparable to that of the battery made from virgin materials. A counter to this argument is the distinct possibility that the recovery of these materials will increase the supply and, as a result, the cost of the materials will decrease, resulting in some lowering of the cost of a battery.

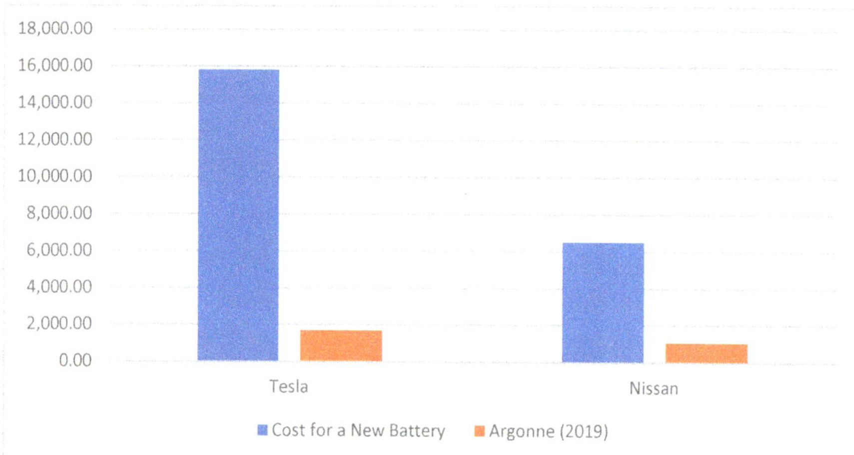

Figure 7. Creating new batteries vs. recycling.

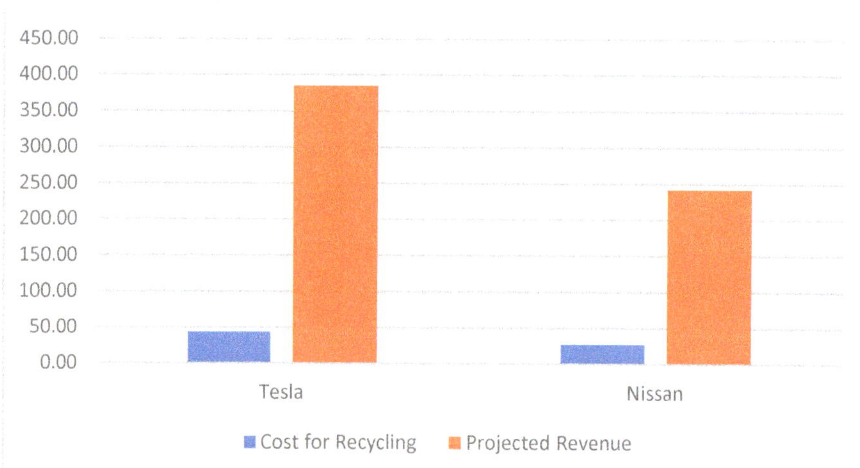

Figure 8. Recycling cost vs. projected revenue.

6. Discussion

The findings presented in this study advance the field of electric vehicle (EV) battery recycling through a detailed comparison of the Life Cycle Assessment (LCA) results obtained using the SimaPro V7 and GREET V2 software, with an emphasis on the environmental and economic implications of recycling nickel–cobalt–manganese (NCM) chemistry batteries.

This analysis is particularly valuable in light of recent research efforts that have similarly employed these tools to assess the sustainability of EV batteries. Notably, the findings presented here reveal a significant reduction in environmental impact and energy consumption through the use of recycled materials when compared to virgin battery production. A similar observation was made by Harper et al. [38] in their comprehensive review on the life cycle environmental impacts of lithium-ion batteries. However, where Harper et al. [38] underscore the challenges in quantifying the specific contributions of battery components to overall environmental degradation, this study leverages the updated databases and methodologies of SimaPro and GREET to offer a more comprehensive analysis related to materials such as nickel, copper, and graphite.

Moreover, the economic analysis presented in this research, highlighting the cost-effectiveness of recycling when it comes to new battery production, aligns with the findings of Gaines and Cuenca [35,39], who demonstrated the potential for considerable cost savings and revenue in battery recycling. Nevertheless, this study takes a deeper look at these economic impacts by integrating a Life Cycle Cost Analysis (LCCA) to quantify the profitability of recycling operations, hence providing a more detailed understanding of the economics within which these processes occur.

In contrast to Wang et al. [40], who highlighted the technical efficiencies of various recycling processes without fully addressing their environmental or economic contexts, this research adopted a more holistic approach. By doing so, it not only reaffirmed the technical feasibility of battery recycling, as demonstrated by Wang et al. [40], but also broadened the discourse with a thorough environmental and economic analysis. This comprehensive perspective is crucial for stakeholders aiming to optimize both the sustainability and profitability of EV battery recycling.

It is also important to contextualize the findings reported in this study within the limitations acknowledged, particularly when it comes to the reliance on outdated databases and the simplifications inherent in economic analyses. These limitations reflect the broader challenges facing LCA studies in achieving both precision and applicability, as highlighted by Notter et al. [41] in their analysis of the environmental impacts of lithium-ion batteries. As such, this study contributes to the ongoing effort to refine LCA methodologies and data sources, offering a steppingstone towards more accurate and applicable guidelines in the context of the sustainability of EV battery recycling.

Also, the results presented in this research validate and extend the findings of existing research on the environmental and economic benefits of EV battery recycling. They also provide new insights through the application of updated LCA tools and methodologies. By doing so, this study not only reinforces the importance of recycling in mitigating the environmental footprint of EV batteries, but also clarifies the potential for recycling as a profit-generating business, thereby supporting the advancement of sustainable practices in EV battery management.

6.1. *Economic Analysis of Recycling Technology*

Material recycling is not only advantageous for the environment, but for the economy and society as well. The activities needed to locate, mine, and extract raw materials can be replaced with the use of recovered materials from used batteries. All of these activities need energy, destroy the surrounding landscape, and pollute the environment. Moreover, the search for raw materials has deep political and social implications, with minerals such as cobalt, copper, and lithium being the root causes for several human rights violations [42].

On the global stage, there is a significant push away from the traditional linear economy, where materials are manufactured, used, and then disposed, whereas with the circular economy, after disposal, the raw materials of a product are extracted, and new batteries are manufactured [43]. In Europe, waste management infrastructure is sufficiently advanced to allow for an almost complete collection of solid wastes [44]. However, after collection, there is a notable drop in the recycling of those collected materials. In fact, an EEA report [45] estimated that even when considering the most optimistic economic scenarios, recycled materials from wastes constitute less than 0.5% of the European GDP, with recovered materials only representing 5% to 15% of the materials used in manufacturing and construction in most EU countries [46].

A total of 5.7 billion tons of material has been consumed by the EU economy to support its population's demand for goods and services in 2013, amounting to approximately EUR 400 billion [43]. The trend of increased consumption is expected to increase as the world population grows, with estimates suggesting an 800% increase in consumption by 2050 when compared to 1990′s level [47]. Consequently, experts have asserted that unless the process in which raw materials are disposed of and recycled changes, the current stock of virgin material that is still unextracted will not be enough to support the needs of future generations [48], with around 100 billion tons of raw materials having entered the economy in 2020 alone [49].

On a more positive note, the transition towards a circular economy is expected to yield USD 500 billion in savings for the European economy [50], and it is estimated that it will create one million jobs in the recycling and remanufacturing industries [51]. From a strategic perspective, adopting the CE approach reduces a country's reliance on material imports, shielding it from potential supply disruptions and price volatility, as was seen during the COVID-19 pandemic [43,52].

LIB recycling fits perfectly in the context of a CE. LIBs are the technology of choice for electric and hybrid vehicles, and they can be re-used in the context of stationary energy storage solutions given that LIBs retain about 80% of their capacity at the end of their first use [53,54]. LIBs experience widespread usage in various private and industrial applications, including commercial electronics such as laptops and smartphones [55,56]. In fact, for EVs alone, the global demand for battery capacity is expected to increase from 120 GWh in 2019 to 1525 GWh in 2030 [57]. In addition, China is positioned as the largest EV market in the world and is the primary contributor to this increasing demand, which was made possible through a combination of market dynamics, governmental policies, and manufacturing capabilities [58]. Hence, a global push towards recycling is necessary.

However, the global recycling rate for LIBs is still less than 5% [59]. To tackle this problem and reduce the world's reliance on virgin materials, the CE approach suggests two strategies: recycling LIBs to recover raw materials such as lithium, cobalt, and manganese and re-using LIBs in stationary energy systems or other applications [60,61]. Additionally, innovative recycling technologies are poised to increase the efficiency of material recovery and reduce waste. Nonetheless, while there is a legislative push by policymakers to encourage circular economy implementation, challenges remain such as material losses and the need for further processing in order to reuse LIBs in different applications [61]. Additionally, there are also important technical and economical hurdles that prevent recycling technologies from achieving high recovery rates, as well as a lack of data with regard to reuse and remanufacturing technologies, which are needed in order to evaluate the usage of LIBs in secondary applications [62,63].

6.2. Political Motivations behind Recycling Adoption

The EV battery market is largely dominated by China in terms of production and recycling, which has sparked concerns by U.S. and European lawmakers, given the importance that LIBs have on the global stage. Indeed, China accounted for about 80% of global LIB production, and its EV sales neared six million in 2022. This dominance is the product of substantial investment by the Chinese government in the development of a reliable infrastructure that supports this level of production, in addition to significant investment in research and development efforts [64]. China is also the global leader when it comes to LIB recycling, with this market being expected to increase from USD 11 billion to USD 18 billion by 2028, which represents approximately the entirety of the recycling market [65].

Moreover, current trends suggest that China's role as a global leader in LIB production and recycling is assured for decades to come, with researchers estimating that China will reach lithium self-sufficiency from recycling by 2059, meaning it is way ahead of competitors such as Europe and the U.S., which are poised to reach self-sufficiency by 2070 [66]. Additionally, researchers found that China will meet its demand for nickel and cobalt by 2045 and 2046, respectively, well before Europe and the U.S. [66,67].

Given the following, Western countries have to make significant efforts in order to overtake China as global leaders in the LIB space. In fact, the Inflation Reduction Act of the U.S., which was signed into law in 2022, includes several provisions that promote local EV recycling initiatives. This has prompted significant investments in companies, such as Ascend Elements and Redwood Materials, with the aims of reaching a closed-loop supply chain and reducing the U.S.'s reliance on Chinese imports [68]. On the other hand, China is taking measures to reinforce its lead, with several policies being implemented that aim to enhance research and set stringent recycling standards, opposing the efforts made by the U.S., which China has accused of being "anti-globalist" [69]. The aggressive implementation of these measures highlights the increasingly political nature of the recycling race, with the level of governmental support playing a crucial role in the trajectory of EV battery production and recycling [70].

6.3. *The Specific Contributions of This Study*

This study substantially advances the existing body of knowledge regarding the secondary use of lithium-ion batteries (LIBs), which, until now, primarily assessed the impacts at isolated environmental, societal, or economic levels. This research takes a novel approach by offering a holistic analysis that considers the combined downstream effects of LIB misuse, marking a pioneering effort in this field. Through the use of advanced analytical software, namely SimaPro V7 and GREET V2, this research quantifies environmental impacts in fifteen distinct categories. This comprehensive quantification not only provides a direct comparison with the environmental costs associated with manufacturing new batteries, but also clearly showcases the considerable environmental advantages of battery recycling.

Additionally, this study explores qualitative impact factors that are often overlooked due to the difficulty in quantifying them. These include noise and thermal pollution, as well as the degradation of soil and water quality, which significantly deteriorate living conditions. Furthermore, it is often the poorest and most marginalized communities that are disproportionately affected by those actions, as they tend to live closer to industrial sites [71], further exacerbating social inequalities.

Economically, this research outlines the viability of recycling as a profitable business, bolstered by current economic policies and geopolitical dynamics. The findings show that recycling not only has substantial economic benefits but also aligns with recent governmental fiscal incentives aimed at supporting domestic raw material production over imports. This aspect of the study underscores the strategic importance of recycling initiatives in strengthening local economies and reducing the dependency on foreign resources.

Overall, this expanded investigation shows a comprehensive picture of the multifaceted impacts of LIB recycling. It equips stakeholders, policymakers, and the general public with a deeper appreciation of the benefits of recycling, encouraging broader support for sustainable practices. By providing a thorough analysis of both the quantifiable and qualitative effects, this research offers valuable insights that can lead to more informed decisions, fostering a more sustainable and equitable approach to battery use and recycling in the technological era.

7. Summary and Conclusions

The world must be ready to effectively handle the influx of end-of-life (EOL) electric vehicle (EV) lithium-ion batteries (LIBs). Effective management is crucial to mitigate the significant environmental and economic consequences associated with improper disposal. Recycling presents an optimal solution, enabling the reuse of materials, thereby reducing both the cost of new batteries produced from recycled LIBs and the emissions associated with raw material extraction. Conducting a Life Cycle Assessment (LCA) provides a detailed analysis of each stage in the battery's lifecycle from extraction to disposal. SimaPro and GREET enabled the evaluation of emissions from both new and recycled batteries despite the differing presentation styles of the two software packages, and both indicated that recycling has reduced levels of environmental impacts compared with the use of new materials.

Additionally, a Life Cycle Cost Assessment (LCCA) revealed the economic viability and commercial potential of recycling. Future research endeavors should include several key initiatives to enhance the understanding and evaluation of EV battery recycling. Hydrometallurgy and pyrometallurgy processes should be incorporated into the GREET software analysis to better assess the environmental impacts of these recycling techniques. Lastly, future research should concentrate on quantifying both the environmental and economic costs associated with producing new batteries using recycled materials. This thorough approach would deepen the insights into the recycling process and assess the practicality and sustainability of using recycled components obtained from alternative processing techniques in second-use battery production.

In the United States, recycling companies have established factories in various states, such as Nevada, which includes companies like Redwood Materials, NV and American Battery Technology Co.; Texas, which includes companies like Ecobat; Massachusetts, which includes companies like Ascend Elements; and New Jersey, which includes companies like Princeton NuEnergy, to name a few. While the topic of this study revolves around the state of New Jersey, states in which recycling companies have factories are making a push towards making their energy mixes more environmentally friendly by increasing their reliance on renewable sources like wind and solar energy and decreasing their reliance on fossil fuels. Ganji et al. [72] denote that the U.S. electric sector is expanding its use of natural gas and renewable energies and moving away from the traditional reliance on coal and oil. The findings in this paper can safely be generalized to other states that house recycling companies; however, future research should still endeavor to study energy mixes in other states primarily to gain better insights on the current state of U.S. infrastructure and its suitability to allow for further transition towards renewables.

Author Contributions: Conceptualization, J.M.; methodology, J.M. and D.W.; validation, J.M., D.W. and G.C.; formal analysis, J.M. and G.C.; investigation, G.C.; writing—original draft preparation, G.C.; writing—review and editing, J.M., D.W. and G.C.; supervision, J.M.; project administration, J.M.; funding acquisition, J.M. All authors have read and agreed to the published version of the manuscript.

Funding: This research was funded by New Jersey Department of Environmental Protection, grant number SHW23-003, titled "Sustainable Management of Rechargeable Batteries used in Electric Vehicles in NJ".

Data Availability Statement: The original contributions presented in the study are included in the article, further inquiries can be directed to the corresponding authors.

Acknowledgments: The contents of this report reflect the views of the authors, who are responsible for the facts and the accuracy of the information presented herein. The contents do not necessarily reflect the views or policies of NJIT or NJDEP. This paper does not constitute a standard, specification, or regulation. The authors wish to acknowledge the efforts of the NJDEP project managers, Riley Kirejevas and Chris Mikulewicz, and the contributions of Sarvagna Malladi, NJIT Provost Highschool Scholar. The AI software ChatGPT was used to rephrase and improve the presentation of the manuscript but not the content.

Conflicts of Interest: The authors declare no conflicts of interest.

References

1. Klöpffer, W.; Grahl, B. *Life Cycle Assessment (LCA): A Guide to Best Practice*; John Wiley & Sons: Hoboken, NJ, USA, 2014.
2. Global Market Insights Inc. U.S. Electric Vehicle Market Size & Share: Forecast Report, 2032. 2023. Available online: https://www.gminsights.com/industry-analysis/us-electric-vehicle-market (accessed on 14 May 2024).
3. Meegoda, J.N.; Malladi, S.; Zayas, I.C. End-of-Life Management of Electric Vehicle Lithium-Ion Batteries in the United States. *Clean Technol.* **2022**, *4*, 1162–1174. [CrossRef]

4. IEA. Trends in Batteries—Global EV Outlook 2023—Analysis. 2023. Available online: https://www.iea.org/reports/global-ev-outlook-2023/trends-in-batteries (accessed on 14 May 2024).
5. Are Electric Car Batteries Bad for the Environment? EVBox. 2023. Available online: https://blog.evbox.com/ev-battery-environmental-impact (accessed on 14 May 2024).
6. The Shocking Effect of Electric Car Battery Disposal on Local and Global Public Health. 2023. Available online: https://energy5.com/the-shocking-effect-of-electric-car-battery-disposal-on-local-and-global-public-health (accessed on 14 May 2024).
7. EIA. The United States Surpassed Two Million On-Road Light-Duty Electric Vehicles in 2021. U.S. Energy Information Administration. 2023. Available online: https://www.eia.gov/todayinenergy/detail.php (accessed on 14 May 2024).
8. Rosenberg, S.; Kurz, L.; Huster, S.; Wehrstein, S.; Kiemel, S.; Schultmann, F.; Reichert, F.; Wörner, R.; Glöser-Chahoud, S. Combining dynamic material flow analysis and life cycle assessment to evaluate environmental benefits of recycling–A case study for direct and hydrometallurgical closed-loop recycling of electric vehicle battery systems. *Resour. Conserv. Recycl.* **2023**, *198*, 107145. [CrossRef]
9. Hoover, Z.; Nägele, F.; Polymeneas, E.; Sahdev, S. How charging in buildings can power up the electric-vehicle industry. *McKinsey* **2021**, *9*, 1–8.
10. Bobba, S.; Mathieux, F.; Ardente, F.; Blengini, G.A.; Cusenza, M.A.; Podias, A.; Pfrang, A. Life Cycle Assessment of repurposed electric vehicle batteries: An adapted method based on modelling energy flows. *J. Energy Storage* **2018**, *19*, 213–225. [CrossRef]
11. Fan, T.; Liang, W.; Guo, W.; Feng, T.; Li, W. Life cycle assessment of electric vehicles' lithium-ion batteries reused for energy storage. *J. Energy Storage* **2023**, *71*, 108126. [CrossRef]
12. Quan, J.; Zhao, S.; Song, D.; Wang, T.; He, W.; Li, G. Comparative life cycle assessment of LFP and NCM batteries including the secondary use and different recycling technologies. *Sci. Total Environ.* **2022**, *819*, 153105. [CrossRef]
13. Gaines, L.; Sullivan, J.; Burnham, A.; Belharouak, I. Life-cycle analysis of production and recycling of lithium ion batteries. *Transp. Res. Rec.* **2011**, *2252*, 57–65. [CrossRef]
14. Picatoste, A.; Justel, D.; Mendoza, J.M.F. Circularity and life cycle environmental impact assessment of batteries for electric vehicles: Industrial challenges, best practices and research guidelines. *Renew. Sustain. Energy Rev.* **2022**, *169*, 112941. [CrossRef]
15. Kotak, Y.; Marchante Fernández, C.; Canals Casals, L.; Kotak, B.S.; Koch, D.; Geisbauer, C.; Trilla, L.; Gómez-Núñez, A.; Schweiger, H.-G. End of electric vehicle batteries: Reuse vs. recycle. *Energies* **2021**, *14*, 2217. [CrossRef]
16. .Meegoda, J.; Charbel, G.; Watts, D. Second Life of Used Lithium-Ion Batteries from Electric Vehicles in the USA. *Environments* **2024**, *11*, 97. [CrossRef]
17. Marchese, D.; Giosuè, C.; Staffolani, A.; Conti, M.; Orcioni, S.; Soavi, F.; Cavalletti, M.; Stipa, P. An Overview of the Sustainable Recycling Processes Used for Lithium-Ion Batteries. *Batteries* **2024**, *10*, 27. [CrossRef]
18. Zanoletti, A.; Carena, E.; Ferrara, C.; Bontempi, E. A Review of Lithium-Ion Battery Recycling: Technologies, Sustainability, and Open Issues. *Batteries* **2024**, *10*, 38. [CrossRef]
19. EIA. U.S. Energy Information Administration—EIA—Independent Statistics and Analysis. 2022. Available online: https://www.eia.gov/state/analysis.php (accessed on 14 May 2024).
20. Kim, H.C.; Wallington, T.J.; Arsenault, R.; Bae, C.; Ahn, S.; Lee, J. Cradle-to-gate emissions from a commercial electric vehicle Li-ion battery: A comparative analysis. *Environ. Sci. Technol.* **2016**, *50*, 7715–7722. [CrossRef] [PubMed]
21. World Bank. The Southern Africa TB in the Mining Sector Initiative. 2017. Available online: https://www.worldbank.org/en/programs/the-southern-africa-tb-in-the-mining-sector-initiative (accessed on 14 May 2024).
22. U.S. Department of Energy. State of New Jersey ENERGY SECTOR RISK PROFILE. 2016. Available online: https://www.energy.gov/sites/prod/files/2016/09/f33/NJ_Energy%20Sector%20Risk%20Profile.pdf (accessed on 14 May 2024).
23. Muller, N.Z.; Mendelsohn, R. *The Air Pollution Emission Experiments and Policy Analysis Model (APEEP) Technical Appendix*; Yale University: New Haven, CT, USA, 2006.
24. AWV. Industrial Noise Pollution: Combatting Common Causes and Effects. 2024. Available online: https://awv.com/blog/industrial-noise-pollution (accessed on 14 May 2024).
25. Taylor, B. Recyclingtoday.com. 2018. Available online: https://www.recyclingtoday.com/article/shredding-granulating-equipment-sound-abatement/ (accessed on 14 May 2024).
26. Mrozik, W.; Rajaeifar, M.A.; Heidrich, O.; Christensen, P. Environmental impacts, pollution sources and pathways of spent lithium-ion batteries. *Energy Environ. Sci.* **2021**, *14*, 6099–6121. [CrossRef]
27. Which Environmental Pollutants Are Produced by Lithium Batteries? Petro Online. 2023. Available online: https://www.petro-online.com/news/fuel-for-thought/13/international-environmental-technology/which-environmental-pollutants-are-produced-by-lithium-batteries/59865 (accessed on 14 May 2024).
28. EPA. The U.S. Recycling System. 2023. Available online: https://www.epa.gov/circulareconomy/us-recycling-system (accessed on 14 May 2024).
29. Peters, J.F.; Baumann, M.; Zimmermann, B.; Braun, J.; Weil, M. The environmental impact of Li-Ion batteries and the role of key parameters–A review. *Renew. Sustain. Energy Rev.* **2017**, *67*, 491–506. [CrossRef]
30. Kurland, S.D. Energy use for GWh-scale lithium-ion battery production. *Environ. Res. Commun.* **2019**, *2*, 012001. [CrossRef]
31. Seltzer, M. A Better Way to Recycle Lithium Batteries Is Coming Soon from This Princeton Startup. Princeton University. 2022. Available online: https://www.princeton.edu/news/2022/03/01/better-way-recycle-lithium-batteries-coming-soon-princeton-startup (accessed on 14 May 2024).

32. Saftner, D.; Heintz, K.; Hershey, R. *Identifying Potential Hydrologic Impacts of Lithium Extraction in Nevada*; Division of Hydrologic Sciences, Desert Research Institute: Reno, NV, USA, 2023.
33. Brugato, T. EPA Clarifies Hazardous Waste Requirements Applicable to Lithium Ion Batteries. Inside Energy & Environment. 2023. Available online: https://www.insideenergyandenvironment.com/2023/05/epa-clarifies-hazardous-waste-requirements-applicable-to-lithium-ion-batteries/ (accessed on 14 May 2024).
34. Recurrent. Recurrent. Electric Car Battery Replacement Costs. 2023. Available online: https://www.recurrentauto.com/research/costs-ev-battery-replacement (accessed on 14 May 2024).
35. Gaines, L.; Cuenca, R. *Costs of Lithium-Ion Batteries for Vehicles (No. ANL/ESD-42)*; Argonne National Lab.: Lemont, IL, USA, 2000.
36. Dai, Q.; Spangenberger, J.; Ahmed, S.; Gaines, L.; Kelly, J.C.; Wang, M. *EverBatt: A Closed-Loop Battery Recycling Cost and Environmental Impacts Model (No. ANL-19/16)*; Argonne National Lab.(ANL): Argonne, IL, USA, 2019.
37. Curran, P. The Economics around Lithium-Ion Battery Recycling Are Strong and Growing. GLG. 2021. Available online: https://glginsights.com/articles/the-economics-around-lithium-ion-battery-recycling-are-strong-and-growing/ (accessed on 14 May 2024).
38. Harper, G.; Sommerville, R.; Kendrick, E.; Driscoll, L.; Slater, P.; Stolkin, R.; Walton, A.; Christensen, P.; Heidrich, O.; Lambert, S.; et al. Recycling lithium-ion batteries from electric vehicles. *Nature* **2019**, *575*, 75–86. [CrossRef] [PubMed]
39. Gaines, L.; Cuenca, R. *Life-Cycle Costs of Lithium-Ion Vehicle Batteries*; SAE Transactions: New York, NY, USA, 2000; pp. 1920–1931.
40. Wang, L.; Wang, X.; Yang, W. Optimal design of electric vehicle battery recycling network–From the perspective of electric vehicle manufacturers. *Appl. Energy* **2020**, *275*, 115328. [CrossRef]
41. Notter, D.A.; Gauch, M.; Widmer, R.; Wager, P.; Stamp, A.; Zah, R.; Althaus, H.J. Contribution of Li-ion batteries to the environmental impact of electric vehicles. *Environ. Sci. Technol.* **2010**, *44*, 6550–6556. [CrossRef]
42. Whitney, T. Conflict minerals, black markets, and transparency: The legislative background of Dodd-Frank Section 1502 and its historical lessons. *J. Hum. Rights* **2015**, *14*, 183–200. [CrossRef]
43. Di Maio, F.; Rem, P.C. A robust indicator for promoting circular economy through recycling. *J. Environ. Prot.* **2015**, *6*, 1095–1104. [CrossRef]
44. Golder Europe EEIG. Implementation of the Landfill Directive in the 15 Member States of the European Union, A Report, ENV. A. 2004, October 2005. Available online: https://ec.europa.eu/environment/pdf/waste/report_a2.pdf (accessed on 14 May 2024).
45. Mazur-Wierzbicka, E. Towards circular economy—A comparative analysis of the countries of the European Union. *Resources* **2021**, *10*, 49. [CrossRef]
46. Hüther, M. *Volkswirtschaftliche Bedeutung der Entsorgungs—Und Rohstoffwirtschaft*; Institut der Deutschen Wirtschaft: Köln, Germany, 2010.
47. Ayres, R.U. Sustainability economics: Where do we stand? *Ecol. Econ.* **2008**, *67*, 281–310. [CrossRef]
48. Gordon, K. Risk assessment in development lending. *Brief. Real Estate Financ.* **2003**, *3*, 7–26. [CrossRef]
49. Fraser, M.; Haigh, L.; Soria, A.C. *The Circularity Gap Report 2023*; Circle Economy Foundation: Amsterdam, The Netherlands, 2023.
50. Papetti, A.; Menghi, R.; Di Domizio, G.; Germani, M.; Marconi, M. Resources value mapping: A method to assess the resource efficiency of manufacturing systems. *Appl. Energy* **2019**, *249*, 326–342. [CrossRef]
51. De Angelis, R.; Howard, M.; Miemczyk, J. Supply chain management and the circular economy: Towards the circular supply chain. *Prod. Plan. Control* **2018**, *29*, 425–437. [CrossRef]
52. Pellegrino, R. Supply chain finance: How do supply chain strategies perform in mitigating supply chain disruption and commodity price volatility risks? *IEEE Eng. Manag. Rev* 2024, *in press*. [CrossRef]
53. Schwieters, T.; Evertz, M.; Mense, M.; Winter, M.; Nowak, S. Lithium loss in the solid electrolyte interphase: Lithium quantification of aged lithium ion battery graphite electrodes by means of laser ablation inductively coupled plasma mass spectrometry and inductively coupled plasma optical emission spectroscopy. *J. Power Sources* **2017**, *356*, 47–55. [CrossRef]
54. Xu, K. Electrolytes and interphases in Li-ion batteries and beyond. *Chem. Rev.* **2014**, *114*, 11503–11618. [CrossRef]
55. Armand, M.; Axmann, P.; Bresser, D.; Copley, M.; Edström, K.; Ekberg, C.; Guyomard, D.; Lestriez, B.; Novák, P.; Petranikova, M.; et al. Lithium-ion batteries–Current state of the art and anticipated developments. *J. Power Sources* **2020**, *479*, 228708. [CrossRef]
56. Schmuch, R.; Wagner, R.; Hörpel, G.; Placke, T.; Winter, M. Performance and cost of materials for lithium-based rechargeable automotive batteries. *Nat. Energy* **2018**, *3*, 267–278. [CrossRef]
57. UCSUSA. Electric Vehicle Batteries. Union of Concerned Scientists. 2021. Available online: https://www.ucsusa.org/resources/ev-battery-recycling (accessed on 14 May 2024).
58. Ou, S.; Lin, Z.; Wu, Z.; Zheng, J.; Lyu, R.; Przesmitzki, S.; He, X. *A Study of China's Explosive Growth in the Plug-In Electric Vehicle Market*; UT Battelle, LLC.: Knoxville, TN, USA, 2017.
59. Choubey, P.K.; Kim, M.S.; Srivastava, R.R.; Lee, J.C.; Lee, J.Y. Advance review on the exploitation of the prominent energy-storage element: Lithium. Part I: From mineral and brine resources. *Miner. Eng.* **2016**, *89*, 119–137. [CrossRef]
60. Shanmugam, V.; Das, O.; Neisiany, R.E.; Babu, K.; Singh, S.; Hedenqvist, M.S.; Berto, F.; Ramakrishna, S. Polymer recycling in additive manufacturing: An opportunity for the circular economy. *Mater. Circ. Econ.* **2020**, *2*, 1–11. [CrossRef]
61. Velázquez-Martínez, O.; Valio, J.; Santasalo-Aarnio, A.; Reuter, M.; Serna-Guerrero, R. A critical review of lithium-ion battery recycling processes from a circular economy perspective. *Batteries* **2019**, *5*, 68. [CrossRef]
62. Ahuja, J.; Dawson, L.; Lee, R. A circular economy for electric vehicle batteries: Driving the change. *J. Prop. Plan. Environ. Law* **2020**, *12*, 235–250. [CrossRef]

63. Fujita, T.; Chen, H.; Wang, K.T.; He, C.L.; Wang, Y.B.; Dodbiba, G.; Wei, Y.Z. Reduction, reuse and recycle of spent Li-ion batteries for automobiles: A review. *Int. J. Miner. Metall. Mater.* **2021**, *28*, 179–192. [CrossRef]
64. Manjong, N.B.; Bach, V.; Usai, L.; Marinova, S.; Burheim, O.S.; Finkbeiner, M.; Strømman, A.H. A comparative assessment of value chain criticality of lithium-ion battery cells. *Sustain. Mater. Technol.* **2023**, *36*, e00614. [CrossRef]
65. EMR. Market Analysis Report. 2022. Available online: https://www.expertmarketresearch.com/reports/battery-recycling-market (accessed on 14 May 2024).
66. Wesselkämper, J.; Dahrendorf, L.; Mauler, L.; Lux, S.; von Delft, S. A battery value chain independent of primary raw materials: Towards circularity in China, Europe and the US. *Resour. Conserv. Recycl.* **2024**, *201*, 107218. [CrossRef]
67. Kendall, A.; Slattery, M.; Dunn, J. End of Life EV Battery Policy Simulator: A Dynamic Systems, Mixed-Methods Approach. National Center for Sustainable Transportation: Davis, CA, USA, 2024.
68. Dunn, J.; Kendall, A.; Slattery, M. Electric vehicle lithium-ion battery recycled content standards for the US–targets, costs, and environmental impacts. *Resour. Conserv. Recycl.* **2022**, *185*, 106488. [CrossRef]
69. Carey, N.; Lienert, P.; Waldersee, V. Dead EV Batteries Turn to Gold with US Incentives. 2023. Available online: https://www.reuters.com/business/autos-transportation/dead-ev-batteries-turn-gold-with-us-incentives-2023-07-21/ (accessed on 16 May 2024).
70. Xiao, Y.; Su, J.; Chen, L. Resonant Acoustic Vibration-Assisted Cathode Stripping for Efficient Recycling of Spent Li-Ion Batteries. *J. Manuf. Sci. Eng.* **2024**, *146*, 044501. [CrossRef]
71. Pasetto, R.; Mattioli, B.; Marsili, D. Environmental justice in industrially contaminated sites. A review of scientific evidence in the WHO European Region. *Int. J. Environ. Res. Public Health* **2019**, *16*, 998. [CrossRef] [PubMed]
72. Ganji, F.; Liu, L.; Fellows, S. Implications of climate change mitigation and socioeconomic development on the US electric power sector. *Energy Clim. Chang.* **2024**, *5*, 100125. [CrossRef]

batteries

MDPI

Article

Comparative Analysis of Lithium-Ion Batteries for Urban Electric/Hybrid Electric Vehicles

Boris Velev [1], Bozhidar Djudzhev [2], Vladimir Dimitrov [3] and Nikolay Hinov [3,*]

1. Institute of Mechanics—Bulgarian Academy of Sciences, 1000 Sofia, Bulgaria; b.velev@imbm.bas.bg
2. Department of Electrical Measurement Systems, Technical University—Sofia, 1000 Sofia, Bulgaria; b.djudjev@tu-sofia.bg
3. Department of Power Electronics, Technical University—Sofia, 1000 Sofia, Bulgaria; dimitrov@tu-sofia.bg

* Correspondence: hinov@tu-sofia.bg; Tel.: +359-29652569

Abstract: This paper presents an experimental comparison of two types of Li-ion battery stacks for low-voltage energy storage in small urban Electric or Hybrid Electric Vehicles (EVs/HEVs). These systems are a combination of lithium battery cells, a battery management system (BMS), and a central control circuit—a lithium energy storage and management system (LESMS). Li-Ion cells are assembled with two different active cathode materials, nickel–cobalt–aluminum (NCA) and lithium iron phosphate (LFP), both with an integrated decentralized BMS. Based on experiments conducted on the two assembled LESMSs, this paper suggests that although LFP batteries have inferior characteristics in terms of energy and power density, they have great capacity for improvement.

Keywords: comparative analysis; lithium-ion batteries; state of charge; state of health; testing; urban EVs/HEVs

1. Introduction

Electromobility is the process of using electric vehicles as a way to reduce emissions of carbon dioxide and other pollutants that are common to traditional vehicles with internal combustion engines. This term generally includes both passenger cars and heavy-duty vehicles that use electric power. The main goal of electromobility is to create a cleaner, more efficient, and sustainable transport system. Urban electric and hybrid vehicles (EVs and HEVs) have become popular choices for modern urban transportation due to their environmental and operational advantages. However, like any technological product, they also have their drawbacks. The following is a summary of their main advantages and disadvantages [1–3]:

1. Advantages:
    - Reduced carbon dioxide emissions—EVs and HEVs emit significantly less or no carbon dioxide into the atmosphere compared to conventional cars, making them more environmentally friendly.
    - Savings on fuel costs—Using electricity as the main or partial source of energy can significantly reduce fuel costs.
    - Low operating costs—Electric motors require less maintenance than conventional internal combustion engines because they have fewer moving parts.
    - Quiet operation—EVs are significantly quieter than vehicles with internal combustion engines, reducing noise pollution in urban environments.
    - Support from public and public funds—Many governments offer tax breaks, subsidies, and other incentives for the purchase and use of EVs and HEVs.
2. Disadvantages:
    - Limited range—Many electric vehicles still have a relatively limited range per charge compared to gasoline-powered cars.

- Charging time—Charging an EV battery can take significantly longer than filling up a traditional car with fuel.
- Insufficient charging infrastructure—Many cities still lack a sufficiently developed infrastructure for the fast and convenient charging of electric vehicles.
- A high initial cost—Despite declining prices over time, EVs and HEVs often have higher purchase prices than conventional cars.
- The environmental issues of batteries and other elements of these vehicles—The production and disposal of lithium-ion batteries bring environmental risks and challenges, including the mining of rare metals and waste management.

Choosing an urban EV or HEV can be influenced by many factors, including personal preference, the availability of charging infrastructure, and government policies.

Lithium-ion batteries are the main source of energy for electric and hybrid vehicles, including those intended for urban use. They have a number of advantages that make them the best choice for this type of transport [4–6]:

- Compactness and lightness—For an urban vehicle, it is important that the battery occupies a minimum amount of space and adds as little weight as possible to the vehicle. Lithium-ion batteries are lighter and smaller in size than other types of batteries, which make them suitable for inclusion in the construction of small and maneuverable city cars.
- Modularity—Batteries can be designed into modular systems, allowing them to be easily added or removed from the vehicle for maintenance purposes or capacity upgrades.
- Thermal management—Urban vehicles often operate in stop/start conditions, which can cause the battery temperature to rise. Effective temperature management systems are critical to maintaining optimal performance and extending battery life.
- Fast and flexible charging—Lithium-ion batteries usually support fast charging, which is important for urban vehicles, as they can be quickly recharged during short interruptions in service. Also, batteries are often compatible with different types of charging stations, offering flexibility in urbanized areas with diverse charging infrastructure.
- Durability and reliability—In an urbanized environment where vehicles are used intensively, it is important that batteries have high durability and can withstand multiple charging cycles without a significant reduction in their efficiency.
- Environmental sustainability—Given the increasing regulations to reduce emissions and increase sustainability, lithium-ion batteries offer an environmentally friendly solution, as they do not require the burning of fuels and are less polluting than traditional car batteries.

These characteristics of lithium-ion batteries make them suitable for use in urban electric and hybrid vehicles, providing them with reliability, efficiency, and flexibility in energy management.

Thus, lithium-ion (Li-Ion) batteries are currently the best energy storage technology for EVs/HEVs and, as such, have been widely investigated in the literature [7,8]. The focus of previous studies ranges from determining the batteries' carbon footprint and energy consumption during production [9–11] to conducting numerous experiments designed to obtain electric equivalent circuit models based on charging/discharging characteristics with different types of current waveforms or impedance measurements. These current profiles are important because the typical characteristics of cells in a datasheet are given for a constant current, which is not the case in EV/HEV applications. The potential advantages and disadvantages of the most popular types of cells for such applications, namely nickel–manganese–cobalt (NMC) and nickel–cobalt–aluminum (NCA), are widely available, and these are the ones that are primarily used in EVs [1,12,13]. There are, however, fewer equivalent studies covering LFP batteries, as these are considered suboptimal both in energy and power density [12].

On the other hand, batteries for hybrid and electric cars differ in several important characteristics depending on the specific requirements and purposes of these vehicles. The main differences between them are as follows [4,7]:

1. Capacity and energy density

HEV: Hybrid car batteries have lower capacity and energy density than electric car batteries. They are designed to assist the internal combustion engine during acceleration, provide short-term electric traction, and store the energy generated during regenerative braking.

EV: The batteries of electric cars have significantly higher capacity and energy density as they must provide the main source of energy for driving the car. They are designed to allow for a longer mileage on a single charge.

2. Size and weight

HEV: Hybrid car batteries are smaller and lighter because they do not need to provide a large amount of energy for an extended period. This helps keep the overall weight of the vehicle lower.

EV: Electric vehicle batteries are larger and heavier due to the greater capacity required to provide a longer range. This results in an increase in the overall weight of the vehicle, which, in turn, necessitates the use of lighter materials in other components of the vehicle to compensate.

3. Type and chemical composition

HEVs: Hybrid electric vehicles use nickel–metal hydride (NiMH) batteries or lithium-ion batteries. NiMH batteries are cheaper and more reliable but have a lower energy density than Li-ion batteries.

EV: Electric vehicles mainly use lithium-ion batteries due to their high energy density, long life, and relatively low weight. Recently, other types of batteries have been developed, such as solid-state batteries, which promise better performance.

4. Charge and Use Cycles

HEV: Hybrid car batteries are used for short periods and are charged and discharged frequently, but in smaller cycles. They must withstand numerous short charge and discharge cycles.

EV: Electric car batteries are used for longer periods and charged less frequently but require longer charging times. They must withstand greater and longer charge and discharge cycles.

5. Terms of use

HEVs: HEVs operate in smaller temperature and load ranges because the battery is not the only source of energy.

EVs: EVs must function efficiently in a wide range of conditions, including extreme temperatures and high loads, as the battery is the main source of energy that drives the vehicle.

Batteries for hybrid and electric cars have specific characteristics that reflect the different requirements and uses of these vehicles. Hybrid batteries are smaller, lighter, and designed for short charge and discharge cycles, while electric car batteries have a larger capacity and are designed to provide long ranges on a single charge and can therefore operate under more severe operating conditions.

Recent developments in battery management system (BMS) software have led to the creation of efficient algorithms for accurate state of charge (SOC) and state of health (SOH) estimations and the balancing of individual cells when part of a pack, as is the case in an LESMS. The balancing methods used are generally based on active and passive balancing techniques, but methods based on a Kalman filter (KF) also exist [14–16]. These studies allowed for the implementation of intelligent BMSs in high-voltage Li-ion packs used in energy storage systems based on NMC or NCA types of batteries. However, there has been a recent trend among all EV manufacturers to try LFP cells in EVs as they have a significantly lower price because they do not require rare earth materials despite having a lower energy density. These cells can be assembled as a stack with a combined energy greater than 100 kWh [17,18]. As this trend is recent, there are not many direct comparisons between implementing an LESMS based on NCA or NMC cells and implementing LFP cells to reduce costs.

These applications do not need to use a high-voltage bus (400–800 V) as they typically have a voltage around 48–96 V, leading to a smaller number of cells and a smaller final volume. To implement such systems, conducting additional research on low-voltage systems is an important step [19,20]. This is the primary motivation for this paper, where direct comparisons are made between an LFP and NCA system with the same energy, working with voltages under 100 V. The experimental data are gathered from real conditions using a mechatronic system that is mounted on a test vehicle [21].

Recent advances in battery technology for urban EVs/HEVs are aimed at improving the energy density, extending the battery life, increasing safety, and reducing costs. These innovations will continue to support the growth and spread of electric and hybrid vehicles, making them more affordable and efficient for consumers. Some of the main innovations and achievements made are as follows [6,9,13]:

1. Solid-State Batteries

Solid-state batteries are one of the most promising new technologies. They use solid electrolytes instead of liquid ones, which increases the energy density and safety of the batteries. These batteries are less prone to overheating, and the possibilities of short circuits are greatly reduced. Examples of progress in this area include developments from companies such as Toyota and QuantumScape.

2. Lithium–sulfur (Li-S) batteries

Lithium–sulfur batteries offer significantly higher energy density than traditional lithium-ion batteries. They use sulfur as the cathode material, which reduces costs and increases capacity. Although there are still challenges related to cycle life and stability, advances in materials science and engineering technology continue.

3. Improvements in electrode materials

The development of new anode and cathode materials, such as silicon anodes and nickel–cobalt–manganese (NCM) cathodes, is helping to increase battery capacity and life. Silicon anodes, for example, can store more lithium ions than traditional graphite anodes.

4. Fast charging

New technologies and methods of accelerated charging allow a significant reduction in battery charging time. An example of this is the development of new electrolytes and improved thermal management methods that allow rapid charging without overheating the batteries.

5. Intelligent Battery Management Systems (BMSs)

Battery management systems are becoming increasingly intelligent, using machine learning algorithms and artificial intelligence to optimize charging and discharging, monitor battery health, and extend battery life. These systems can predict and prevent potential problems, thereby increasing the safety and efficiency of batteries.

6. Recycling and sustainability

The sustainable production and recycling of batteries is becoming increasingly important. New recycling methods enable the extraction and reuse of valuable metals, such as lithium, cobalt, and nickel, reducing dependence on raw material extraction and minimizing the environmental footprint.

7. Thermal management systems

Improvements in thermal management systems help maintain the optimal operating temperature of batteries, increasing their efficiency and life. The use of new cooling materials and passive and active cooling methods reduces the risk of overheating and increases safety.

2. Goals

The goal of this study is to conduct basic tests on low-voltage LESMSs for use in low-range urban EVs/HEVs. Two experimental systems are built—one using NCA cells and the other using LFP cells. These are integrated in a test vehicle to collect data during real driving cycles. The collected data allow us to make direct comparisons between the two systems and to draw conclusions about their potential for uses in low-cost urban EVs/HEVs and in other possible applications such as off-grid PV systems. In this way, we can fill the gap in the literature on the subject related to the testing of energy storage elements in real operating conditions related to their use for urban transport vehicles.

3. Materials and Methods

For a proper comparison between NCA and LFP cells, the characteristics of both cells themselves and the assembled battery packs must be examined. These should be established by using appropriate instrumentation to assemble and provide long-lasting energy storage and management systems (LESMSs) with high power, safe low voltage and operation, zero carbon emissions, and low costs [22,23].

3.1. Types of Li-Ion Cells Used in EVs/HEVs

Only two types of Li-ion cell chemistries have stood the test of time and are widely used for energy storage in EVs/HEVs [4,13]:

- Lithium–nickel–manganese–cobalt (NMC). The active cathode material is $LiCoNiMnO_2$. The nominal voltage of one cell is in the range of 3.6–3.7 V [24].
- Lithium–cobalt–nickel–aluminum (NCA). The active cathode material is $LiCoNiAlO_2$. The nominal voltage is in the range of 3.6–3.7 V [25].

These are lithium cells that are based on cobalt and have high energy density, and as such, they are widely used in modern EV/HEV battery stacks. Their charge characteristics are almost identical. They have a relatively low flash point at 215 °C and are inclined to the "thermal runaway" effect, which warrants additional protection circuits [26–28]. The NCA-type cell has a higher relative energy density in comparison to the NMC-type cell, but it is more expensive.

Relatively recently, some vehicle manufacturers tried using an LFP cell for energy storage:

- Lithium–iron–phosphate (LFP). The active cathode material is $LiFePO_4$. The charging voltage is in the range of 2.1–3.65 V, but it can be up to 3.9 V if doped with yttrium—whose active cathode material is $LiFeYPO_4$. The nominal voltage is in the range of 3.2–3.35 V [24]. These cells have a flash point of 270 °C and, as such, are harder to ignite than NMC or NCA in the case of improper charging or mechanical damage.

The anode material of all cells is graphite.

A comparison of the cells used is given in Table 1. It should be noted that due to the smaller capacity and energy of an individual NCA cell, a comparison is made between 26 cells in parallel. This way, the cells are compared with relatively the same energy. The data are from the manufacturers [24,29].

Table 1. The key parameters of the used cells.

Cell Chemistry	Capacity C [Ah]	Dimensions of One Cell [mm]	Nominal Voltage [V]	Weight [kg]	Energy [Wh]
LFP	100	150 × 180 × 70	3.3	3	330
NCA	91 (26p)	18 × 65	3.65	1.3	332

3.2. Stack Assembly

The battery stack is made using serial combinations of cells. Parameters of individual cells are given in Table 1.

- Li-ion battery consisting of 24 series connected prismatic LFP cells with chemistry

$LiFeYPO_4$ has a nominal size of 150 × 180 × 70 mm, a capacity 100 Ah, and a nominal voltage of 3.33 V. The open-circuit voltage (OCV) of the battery is −79.2 V, and the energy of the stack is 7.92 kWh [29].

- Li-ion battery stack consisting of 676 cylindrical NCA cells with chemistry

For $LiNiCoAlO_2$, the nominal size of each cell is 18 × 65 mm, the capacity 3.5 Ah, and the nominal voltage 3.65 V [2]. The cells are combined in 26p (26 cells in parallel) with OCV − 3.65 V and a capacity of ~91 Ah. With this combination, one 26P combination NCA cells is approximately equal in capacity to one LFP cell. The final assembly is 26P × 26S (26 of the individual parallel cells are combined in series), the OCV is 96.4 V, and the energy is 8 kWh. A 10 A fuse is connected in series with each cell for additional protection and to reduce the chance of "thermal runaway" of the battery in the case of a fire [19,28,30]. Both cell stacks have approximately the same energy −LFP − 7.92 kWh and NCA − 8 kWh. The data of the used battery are summarized in Table 2.

Table 2. Battery stack comparison.

Cell Chemistry	Capacity @ 1C [Ah]	Series Connections	Nominal Total	Energy [kWh]	Total Cell Only Weight [kg]
LFP	100	24	79.2	7.92	72
NCA	91 (26p)	26	96.4	8	34.8

3.3. Experimental Vehicle

The assembled cells, alongside an intelligent BMS, are integrated in a test vehicle. The vehicle has an integrated mechatronic system for collecting the data during the charge/discharge process of the battery in a real environment. This system was thoroughly presented in [21]. Figure 1a,b shows how the cells are integrated in the back seat of the vehicle.

(**a**) (**b**)

Figure 1. Test vehicle with assembled battery stacks for conducting test. (**a**) LFP type; (**b**) NCA type.

The mechatronic system has various types of data acquisition devices and sensors; for impedance measurements, the data acquisition device used was LabJack—U3 (LabJack Corp., Lakewood, CO, USA) with a programmable battery charger and indicator blocks. The system has a graphical user interface (GUI) for the automatic acquisition and analysis of key data while driving. This interface was developed in LabVIEW 2024 Q1 [21], and a block diagram is shown in Figure 2.

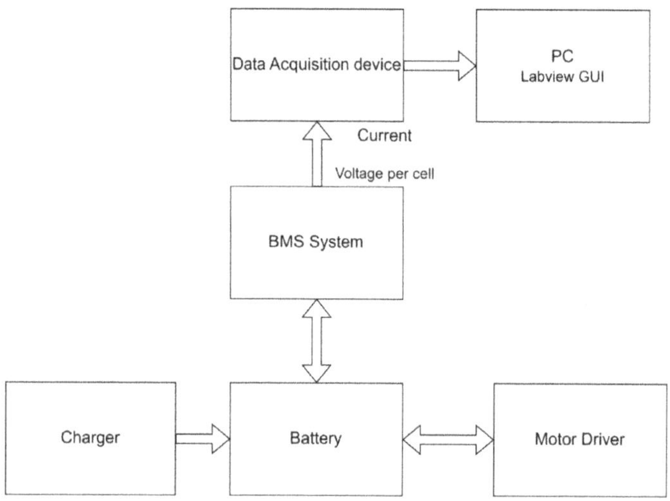

Figure 2. A block diagram of the developed system.

3.4. *Intelligent Module System for Battery Control—BMS*

There are three main objectives that are common for every type of BMS:

- To protect the individual cells from damage due to over discharge or overcharge.
- To prolong the life of the battery by optimally equilibrizing the charge in each cell of the battery.
- To keep the battery in a state which allows for its optimal use.

The developed BMS measures the voltage and current of each cell during operation. The modules for each cell are decentralized and provide serial communication to a master device. When the battery is not in use, the BMS enters a low-power economy mode. When charging, the cell balancing function is performed automatically. The schematic diagram of the decentralized BMS is shown in Figure 3a. On each LFP-type cell (or on each NCA type module) of the battery packs, a local BMS module is installed, which consists of one electronic control unit (ECU) and one passive equalizer EQU, where the ECU controls the EQU and communicates with the central microcontroller of the BMS. Local BMS modules send signals to the BMS control panel on a central processor or computer. Active balancing of the cells is performed with an intelligent module with active EQUs, which is connected to the central processor, to a smartphone, or to a computer [31]. The algorithm of the modular BMS is designed with the main functions shown below.

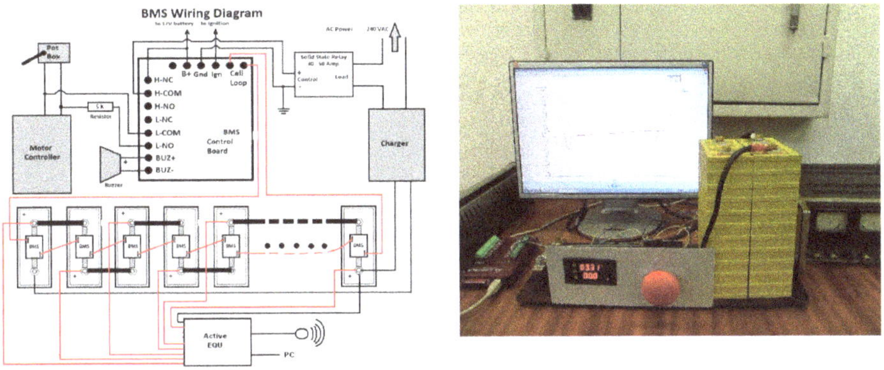

(**a**) (**b**)

Figure 3. (**a**) A block diagram of the intelligent BMS. (**b**) The experimental setup for measuring the internal resistance of LPF cells.

- Voltage measurement. Each ECU in the local BMS modules is shown in Figure 3a. Circuits are used for precise voltage measurement, and they are sent through an interface to the central microprocessor in the control panel.
- Current and state of charge (SOC) measurement. A resistive shunt measures the current though the cells. A counter counts the charge accumulated/extracted for a certain time interval, which is known as CC (Coulomb counting) [27,32,33]. The SOC measured in this way is obtained by very accurately integrating the measured current in the circuit to avoid many of the errors. The algorithm used for determining the current to actively balance the battery packs uses Formula (1):

$$IB\ [A] = C\ [Ah]/t\ [h], \tag{1}$$

where IB—balancing current; C—battery capacity; t—time for gross balancing.

As with the voltage measurement from the local modules, the current measurement is pre-calibrated and periodically automatically corrected. The *SOC* is defined as the percentage of capacity available inside the battery [15,27]. The ratio of available capacity to maximum capacity can be expressed as the *SOC*, which is calculated using Formula (2):

$$SOC = 1 - \frac{\int idt}{C_n} \tag{2}$$

where i—the measured current depending on the gross balancing time t; C_n—the internal capacity of the battery.

An evaluation of the *SOC* is carried out according to the dynamic CC method (Coulomb counting—ΔQ) [15,32,34]. For example, if the initial state is SOC_0, using 1 A current to charge the battery for t hours, they will add I·t Ah of charge to the battery. Also, if the battery capacity is C, then the final *SOC* can be calculated using Formula (3):

$$SOC_t = SOC_0 + \frac{I \cdot t}{C} \tag{3}$$

The charge level is measured by integrating the current I over time t:

$$\Delta SOC = \frac{\Delta Q}{C} = \frac{\int_{t1}^{t2} Idt}{C} \tag{4}$$

Therefore, this method calculates the *SOC* based on the cell current and capacity over time *t*, i.e.,

$$SOC = f(I, t, C) \quad (5)$$

Battery capacity is calculated by integrating the current flowing to or from the battery over time (4). Although the Coulomb counting method is widely used, the error caused by the current or time deviation of the measurements is usually cumulative and can grow to be very large, unless there is a way to recalibrate the *SOC* regularly. This can be carried out, for example, by resetting the *SOC* to 100% when the battery is fully charged. As seen in the tests visualized in Figure 4b, in the charge/discharge cycle of the LFP, there is a capacity deviation of up to 1.2 Ah in the calculated value. At a higher discharge current, the difference is even greater. It is necessary to calculate the capacitance measurement error according to Formula (4). An inaccurate initial time base cannot estimate the difference between the charge/discharge capacity of the cell within a single test cycle because the discharge/charge capacity does not equalize. LabVIEW battery current measurement errors show a maximum error compensation of ΔI = 50 mA. The total time $t_{discharge}$ to discharge is 4 h, but the total time t_{charge} to charge is 4.7 h due to the decreasing current towards the end of the charge. Pause times are subtracted from the above values because current is not measured during pauses. The total difference in the measured capacitance C_{calc}, compensated by these errors, can now be calculated as follows:

$$C_{calc} = 4\text{ h} \times 50\text{ mA} + 4.7\text{ h} \times 50\text{ mA} = 4.50 + 4.7 \times 50 = 0.435\text{Ah}, \quad (6)$$

Therefore, systematic measurement errors alone cannot explain this difference of 0.43 Ah between charge and discharge. Other factors, for example, differences in test conditions such as the temperature, mechanical connections, etc., should be considered as causes of this random error. To remove these errors in the *SOC* determination algorithm for LFP-type cells, a correction factor (k = 2.8) is introduced to periodically correct the capacity measurement. We assume that $C_{calc} = \pm\Delta C_{calc}\, k$.

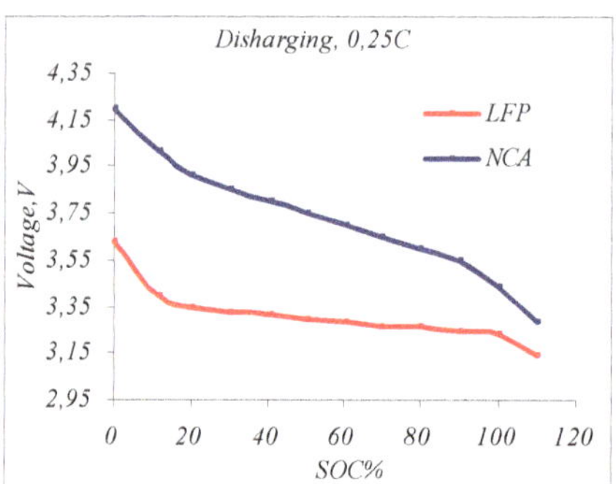

(**a**)

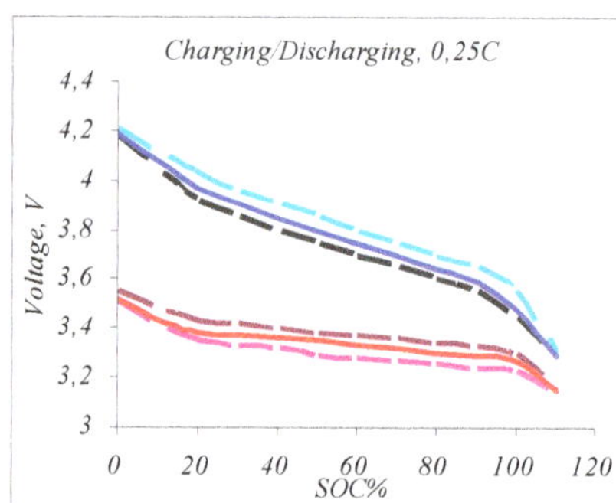

(**b**)

Figure 4. (**a**) Discharge curve of a large LFP cell and a small NCA cell, (**b**) Charge discharge with pause for large LFP and small NCA cell.

Furthermore, in functional tests, voltages and currents depending on the SOC are continuously measured without pauses; periodically, the *SOC* is corrected with C_{calc}; and the accuracy error for *SOC* determination is about 2%. Similar accuracy calculations are made with NCA-type cells where k = 1.5. For each BMS function described above, graphical algorithms that are necessary for the system software have been developed. The integrated current and voltage measurement signals are fed into the BMS control panel via a graphical user interface (GUI) for processing and analysis.

- Cell balancing. The purpose of cell balancing is to increase the battery's performance without overcharging or over-discharging it. Its essence is to make the SOC levels of the cells closer to each other. In the developed smart BMS, a combined passive and active cell balancing method is selected with a decentralized arrangement of the modular BMS, as shown in Figure 3a. The purpose of using passive equalizers is to achieve fast SOC equalization during charging by applying charge shunting to the fastest charging cells until their voltages are within ±0.01 V of the normal cells' OCV voltage, V. Balancing with active EQUs works during battery charging and discharging as well as in standby mode, actively transferring a current of up to 1 A from the higher voltage cells to the lower voltage cells until the voltages equalize. Through the GUI, modules with local passive and active EQUs signal the problems to the central BMS module, where a decision will be made about when to start balancing with the EQU and accurately report the state of charge (*SOC*) [14,16,17,35].

4. Methods for Testing Battery Systems (LESMSs)

4.1. Preliminary Tests

Preliminary tests are conducted to obtain data on the electrochemistry behavior of the corresponding Li-ion cells during discharging and charging. The findings from these initial tests are then used to develop qualification tests for specific purposes in the battery pack, for example, to detect a "weak" cell, to accurately determine the SOC, to determine the current balance of cells in battery packs, etc.

Identification of weak cells. This is determined by measuring the internal resistance. Figure 3b shows a photo of the device, which is used to measure the resistance of prismatic LFP cells. For the test, a pulsed current of 1.5 C is applied for a time of 2 s. Using a 2 s time for the pulse is sufficient to allow the cell voltage to relax, but it has a negligible effect on the SOC. In this case, the pulse current for LFP cells is 150 A, and the current for NCA cells is 5 A, respectively. The current from the battery is measured by a Hall effect current sensor with the current output. The ratio between the secondary current and primary current is 1:5000. Putting a resistor with the *Rc* value across its output and measuring its voltage, $V_{current}$, leads to Formula (7).

$$\frac{1}{5000} = (\text{ratio}) = \frac{V_{current}/Rc}{\text{I}} \tag{7}$$

- The internal resistance of the LFP is calculated according to Formula (8) [4,36]:

$$R_{in} = \frac{\Delta V_{cell}}{\text{I}}, \text{I} = 150 \text{ A} \tag{8}$$

where *Rc*—the resistance of a resistor to form a pulsed current; ΔV_{cell}—open-circuit cell voltage (OCV); $V_{current}$—impulse current voltage across *Rc*; *I*—impulse current thoroughout the battery; R_{in}—the internal resistance of the battery. The ratio is the gain factor of the Hall sensor used for the current measurements.

Figure 5a,b shows the results of the measured voltage drop ΔV_{cell} for two different LFP and NCA cells during the internal resistance test.

- Discharge characteristics of NCA and LFP cells. One fully charged LFP and NCA.

Charged cells from LFP and NCA are subjected to a continuous discharge with a current of 0.25 C (LFP—with 25 A; NCA—with 0.85 A), where C is the battery capacity taken from the manufacturer's data. This discharge allows for the identification of a usable voltage range, the SOC lookup table generation based on voltage, and the comparison to typical discharge curves given by the manufacturer. The results are shown in Figure 4a.

- Full charge/discharge cycle with 0.25 C. Every charge cycle could be separated into 4 different phases: discharge, pauses, charging, and pauses. By applying the LabVIEW graphical program, we can control the phase change boundary condition.

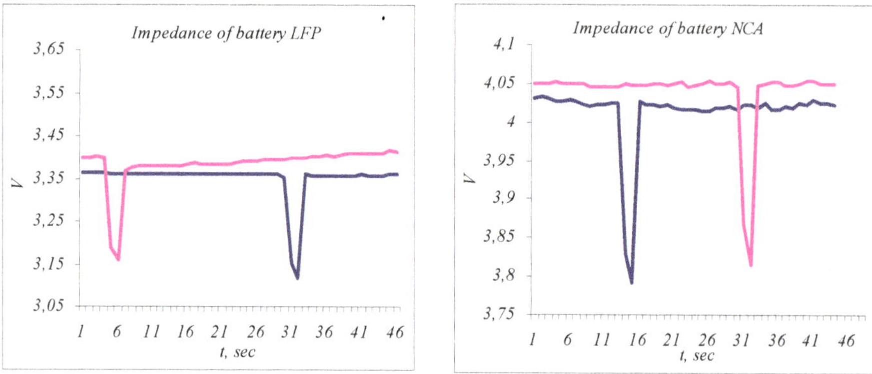

(a) (b)

Figure 5. (**a**) The pulses of the V_{cell} during the internal resistance test of the two LFP cells. (**b**) The pulses of V_{cell} during the internal resistance test of the two NCA cells.

In Figure 4b, cell voltage curves during a preliminary test cycle with a 25 A charge/discharge current with pauses of a large LFP cell and a 0.85 A charge/discharge current of a small NCA cell are shown depending on the SOC.

Determining the SOC of LFP-type stacks. As can be seen in Figure 4b, the voltage slope of LFP cells is very small. This fact complicates the software algorithm's ability to accurately determine the SOC, unlike NCA type cells, where the curve is sufficiently steep, and the SOC is easily determined by measuring the voltage alone. Therefore, it is necessary to test the LFP to investigate the accuracy of determining the SOC rate measured by the capacity during the full discharge/charge cycle test according to Formula (4).

4.2. Qualification Tests

Qualification tests are designed to determine whether an LESMS consisting of a battery pack and BMS is fit for the application. They are conducted after the mandatory preliminary procedures and tests of the respective cells. The LESMS must be tested with a compatible charger so that we are confident that the batteries have not been inadvertently overcharged or damaged [21]. The technique of the qualification tests is close to the real operating conditions because the LESMSs are tested together for compatibility with the main components of EVs/HEVs—electric motors, a mechanical transmission, and running wheels [21]. The dynamic tests are carried out with the mechatronic system in real conditions (on a test site) and in busy city traffic (Figure 6a,b). The main qualification tests are the following:

- Initial charging of the system. Only after the cells are balanced on the same voltage level and after checking the internal resistance and technical characteristics is it possible to assemble them together with the BMS into a functional LESMS.
- Constant current discharge. Fully charged LESMSs undergo a constant current discharge with current I~0.25 C depending on the SOC, where C is the capacity of the LESMS—in the case for both systems, the current is, on average, about 20 A. They are discharged to establish the characteristics of the discharge curves and the operating zones of the systems (Figure 7a).
- Weak cell identification in LFP or weak module in NCA. With the BMS, the general EQU operations of LFP or NCA cells can be counted, and this information can be used to identify the weak cell (module). A weak cell always requires more EQU operations regardless of whether it is in the process of charging or discharging. The LESMS is only loaded to the levels of the weak cell LFP (or the weak module NCA, respectively) to prevent overcharging and degradation.

- Testing under a constant load. A test is carried out to verify that the LESMS can deliver the specified power under a load when required. The load is usually representative of the expected conditions under which the system may be used (Figure 7b).
- Testing under a dynamic load. The tests for cyclic and dynamically changing loads are performed with the test vehicle. For simplicity in testing, no recovery is applied. The operation of a "pure" electric car under different road conditions is simulated. For cyclic loading, the test car's movement on a test site with simulated stopping and starting actions at a traffic light at a certain distance is carried out in order to establish the suitability of the components for application in joint operation (Figure 6a). Figure 6b presents a graph of the current and voltage under a dynamic load of LESMS at a 50 km mileage with a maximum speed of 60 km/h in real urban conditions typical of an urban EV/HEV.

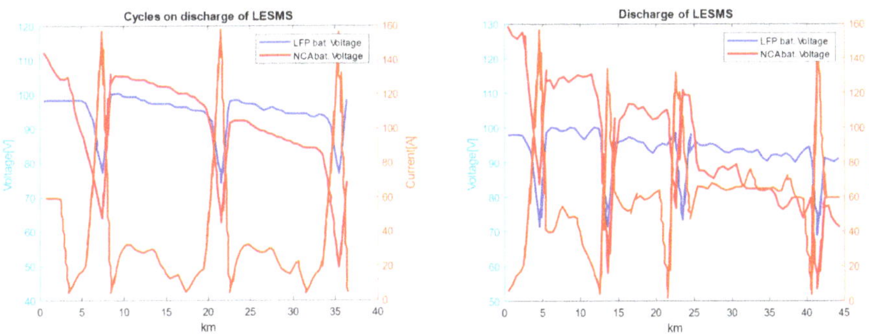

(a) (b)

Figure 6. (**a**) The current and voltage under cyclic discharge of the LESMS for LFP and NCA. (**b**) The current and voltage of the system when using LFP and NCA under a max speed of 60 km/h.

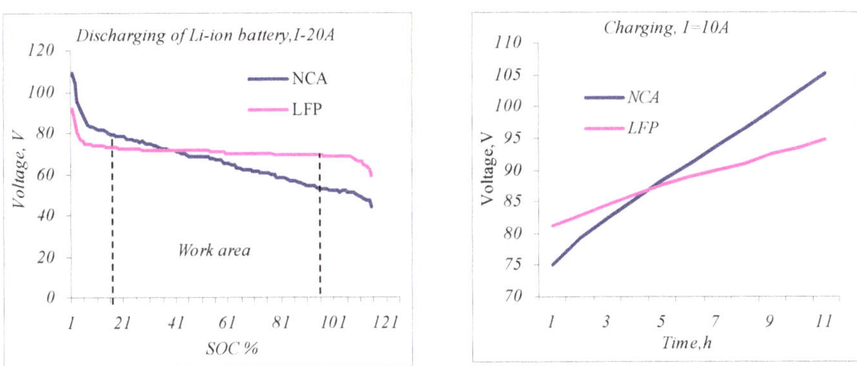

(a) (b)

Figure 7. (**a**) Charging LESMS with constant current of 20 A. (**b**) Discharging LESMS with constant current of 10 A in working voltage area.

5. Experimental Results and Discussion

5.1. The Advantages and Disadvantages of the System

Compared to standard BMSs, the new modular system provides dramatic reductions in cost, size, and weight, all of which are critical for budget and urban EV/HEV applications. Another advantage is that decentralized BMS modules allow the system to be applied to a large number of cells (over 32 cells connected in series) that are easily repaired or replaced, unlike complex BMSs of high-voltage systems for standard EVs. A disadvantage is the vulnerability of this BMS to moisture, dust, and short-circuiting, which is why the entire LESMS needs to be sealed to increase durability and safety. Recently, pressurized competitive intelligent BMSs [32] have appeared on the market, but unfortunately, they still cannot offer the management of series-connected cells above 24 cells (24 S is up to 72 V), which is insufficient for ESBMSs for urban EV/HEV applications, where the control of 32 cells connected in series (32 S to 96 V) is required.

5.2. Results and Discussion of Preliminary Testing

- Weak cell identification. Figure 5a shows the voltage ΔVcell of two fully charged LFP cells, and Figure 5b shows the same ΔVcell of two fully charged NCA cells during the internal impedance measurement according to Formulas (7) and (8). In Figure 5a, the slower OCV relaxation is seen in one of the LFP-type cells; therefore, it has a higher internal resistance. This cell was identified as being weaker than the normal LFP cell. In the two tested cells of the NCA type, the ΔVcell pulses and the voltage, V, at OCV are the same; therefore, both cells have a normal capacity (Figure 5b).
- Discharge and full charge/discharge cycle of LFP and NCA cells. Figure 5a shows the graphs of the discharge voltage versus the SOC. The discharge graph of the small cylindrical NCA cell type is marked with a blue line, and the discharge graph of a large prismatic LFP cell type is constructed with a red line. The voltage difference of the NCA cell at the end and the beginning of the working area is 0.65 V, and the voltage difference of the LFP cell is 0.25 V. Therefore, the electrochemical characteristics of the two cells differ. For example, estimating the state of charge in NCA-type cells becomes much easier by measuring only the voltage at the OCV, while in LFP-type cells, due to the small slope of the discharge curve, estimating the SOC by only measuring the voltage is not a precise enough method. Therefore, in the SOC determination algorithms, for LFP-type cells, it is also necessary to measure the amount of electricity Q (CC) or apply the Kalman filter [15,27,33,35]. In Figure 4b, LFP-type and NCA-type cell voltages are shown as a function of the SOC during the charge/discharge process with 0.25 C current pauses. This allows for a direct comparison of the voltage levels between charge and discharge at the same states of charge. The charging and discharging times are stopped for a short time after every 10% change in capacity to measure and analyze the stress relaxation at the OCV. The upper dashed curves of the cycles represent charging (light blue for the NCA cell and brown for the LFP cell), and the lower dashed curves (black for NCA and light red for LFP, respectively) represent their discharge with 1 min cut-off pauses when the OCV is read. Cell voltage measurements during the test cycle show that the cells do not reach equilibrium at the end of the pauses if the response step has a pause of 1 min, which is very small. Therefore, the equilibrium voltage at the OCV cannot be directly measured at this pause. For full relaxation to the OCV level (equilibrium voltage), longer pauses are required, especially at higher charge/discharge currents and at large LFP cell formats (of the order of 10–15 min). This makes the test difficult and pointless because it extends the time taken to conduct the test cycle to several days. It is necessary to speed up the execution time of the cycle. For this purpose, the continuous averages represent an interpolation between the voltage levels at the end of the 1 min breaks. The blue and red solid curves in the center of the cycles are the averages of this region. Initially, the true equilibrium is assumed to be the OCV. We assume that this approach of estimating the OCV by calculating the average value of the voltage is accurate enough for the charge/discharge cycle with a current in the 0.25–1 C range, where the current most often varies in small city EV/HEV applications; therefore, the continuous curves represent the actual discharge/charge characteristics of the respective Li-ion battery types, and no breaks are required during the test.

5.3. Results and Discussion on Qualification Testing

- The constant current discharge of the LESMS. Fully charged LESMS stacks are discharged with a constant current of I~0.25 C—this equals 20 A, on average, for both cells. In Figure 7a, the discharge curves of the two LESMSs are shown as a function of the SOC. The blue line is the discharge curve of the NCA-type LESMS, and the red line is the discharge curve of the LFP-type system. The visualization of Figure 7a makes it possible to determine the OCV voltage operating zones of the two systems required for the following functional tests. The operating range of the SOC should be between 15% and 90% for NCA and between 10 and 100% for LFP.
- Constant load tests. For this test, the LESMS is loaded with a constant load to verify its use for the application. The LESMSs are charged and then discharged with a constant current of 10 A ranging from 15% to 90% SOC. In Figure 7b, the charging diagram of the two LESMSs is given in the operating range of 15% to 90% SOC for a time of 10 h, as determined by the control system. The visualization shows the higher energy efficiency and steeper curve of the LESMS based on NCA cells (the blue line). Therefore, the NCA system was found to have a higher specific energy of up to 260 Wh/kg, and the LFP-type system was found to have a specific energy of 90 Wh/kg [23,36].
- Cyclic load testing. Figure 6a shows the results of a few cycles with the dynamic discharge of the system composed of LFP and NCA cells. A diagram of the changes in the current and voltage of the two systems under cyclic loads for a certain distance traveled (on a test site) is presented. In Figure 6b, the variations in the current and voltage of the LFP- and NCA-type LESMSs under a load in real urban conditions at a distance of 50 km with a maximum speed of 60 km/h are presented. The currents and voltages are recorded directly from the motor controller via a graphical user interface (GUI) and the LabVIEW graphical program.

As seen in the functional dynamic testing results presented in Figure 6a,b showing the current and voltage variations at a test site and under real urban conditions with an LFP-type LESMS and NCA-type system, it was found that the general thermal and electrical condition of batteries, BMSs, electric motors, and controllers are normal and compatible in different test cycles and there is no difference between them.

5.4. Comparative Analysis

- A comparison of the characteristics of the two types of LESMSs (volume, geometric dimensions, costs, and prices of components and systems). At approximately the same power (~8 kWh), the geometric dimensions of the LFP-type LESMS are approximately twice the size and weight of those of the NCA-type LESMS; therefore, it has twice the energy density and energy efficiency. The assembly costs and the cost of the components (except for the cost of the Li-ion cells) and BMS are about the same for both LESMSs. Recently, in view of the high current price increase for cobalt, nickel, and manganese LFP-type cells, regardless of their lower energy efficiency and complex SOC determination algorithm, they are also starting to be implemented in EV/NEV applications [18,37]. Due to reasons of a modern geopolitical nature, the metals used to manufacture NMC and NCA cells—nickel, cobalt, manganese, and aluminum—have risen sharply in price, and from there, the prices of high-voltage battery systems based on nickel and cobalt have risen sharply [23]. Therefore, for example, "Tesla has announced that it is switching to LFP batteries for its standard range models such as Model 3 and Model Y" [18,37]. The main arguments for the implementation of LFP batteries are the availability of cheap cathode material and the competitive price [23]. Phosphate is much easier to obtain and, for now, reliably cheaper. This means that iron phosphate LESMSs can be much cheaper than cobalt-type NMS and NCA systems and, in addition to solar PV systems, they can successfully displace them in low-cost and urban EV/HEV applications [18,20].

- A comparison based on the preliminary test. It was found that there are significant differences in the electrochemical characteristics of the tested cells, for example, the operating area of the discharge/charge characteristic curve of the NCA-type cells has a greater slope than the almost flat characteristic curve in the operating area of the LFP-type cell (Figure 4a). This means that the SOC estimation of the NCA-type cell is more accurate than that of the LFP-type cell. Correctly calculating the remaining battery capacity is significantly more difficult with LFP cells due to their flatter discharge/charge voltage characteristic. For this reason, it is recommended that LFP-type cells are periodically charged to a 100% SOC so that the BMS can accurately calculate the battery capacity by measuring the amount of electricity, Q, through the remaining capacity, CC, rather than relying solely on the voltage measurement. For NCA-type cells, the voltage measurement is sufficient to estimate the SOC and the remaining capacity. From tests to determine the internal resistance of the cells, it was found that at a 100% SOC, the normal internal resistance of the tested LFP cells was 0.39Ω, and that of the small NCA cells was 0.035Ω. NCA cells have been found to last for 500 to 1000 full charge/discharge cycles, while LFP cells last for 2000 to 6000 full cycles without degrading [24,29]. A disadvantage of LFP cells is their low energy efficiency compared to NCA cells and the difficulty in determining their state of charge (SOC). These problems can be solved during the development and implementation of intelligent control systems with a more complex algorithm, including, in addition to the voltage measurement, the integration of the amount of electricity (Q) through the CC or through an algorithm based on the Kalman filter (KF) [15]. Cells with a higher OCV voltage that reached full charge faster were also found to have higher internal resistance or not be well balanced.
- A comparison based on real exploration tests. The voltage of a no-load current (OCV) of an LESMS is not a reliable measure of its ability to deliver a current. As the battery ages or degrades due to improper use, its internal resistance increases. This reduces the ability to accept and hold a charge, but the OCV idle voltage will still show as normal despite the reduced battery capacity. Comparing the actual internal resistance to the internal resistance of a new battery will show any deterioration in battery functionality; therefore, the internal resistance should be measured and compared periodically.

The correct calculation of an LESMS's residual capacity is significantly more difficult with an LFP cell type due to its flatter discharge voltage–remaining charge curve. For this reason, it is recommended that they are regularly charged to a 100% SOC so that the BMS can calculate the remaining capacity by measuring the amount of electricity through the CC rather than relying solely on the voltage measurement.

The SOC determination accuracy with the developed intelligent BMS in both LESMSs is 2.5%, which is very good for EVs and good for HEVs, where an additional error is introduced by the recuperation when the electric motor works as a generator and charges the battery when driving at work with the internal combustion engine (ICE). This accuracy is achieved by frequently updating the SOC value by the BMS, especially in LFP, to avoid the accumulated error from dynamic operation in real conditions.

Fast charging may reduce the maximum capacity, but this mainly depends on the BMS algorithm and the type of LESMS cells. In the systems tested, especially in the case of the LESMS with LFP-type cells, fast charging to an 80% SOC in 30 min is not possible because forced cooling is required due to the large heat release. The reason for this is that the electrodes of LFP-type prismatic cells are embedded in a heat-insulating plastic housing, which is difficult to cool. With the LESMS with NCA-type cylindrical cells, fast charging is possible because one electrode (anode) is the metal body itself, which can easily be cooled when assembled with a suitable cooling system.

- Suggestions for integration. Recently, low-voltage LESMSs with LFP-type cells, in addition to a solar PV, can also be successfully implemented in low-budget urban vehicles—such as rickshaws, micro EVs, or HEVs—in the conversion of vintage cars weighing up to 750 kg (Citroen 2CV, Renault 4, Suzuki Alto, Trabant 601, etc.). The LESMS with an NCA type cell is recommended for mini urban EVs, and if there is a need for fast charging from charging stations, it can be assembled with a cooling system.

6. Conclusions

The functional testing of the two LESMSs confirmed their suitability for operation in real-world conditions, where they showed equally good operational performance.

LESMSs with cells alloyed with nickel, cobalt, manganese, and aluminum (NCA and NMC types) are at the limits of their refinement and development. The goals of developing LESMSs to acquire more power and achieve a lower price (the goal is to achieve a price lower than USD 100 per kWh) will not be achieved soon. With the envisioned larger market share of electric cars and solar PV systems, the development of NCA and NMC cells will be blocked by environmental requirements for disposal due to the health hazards of the materials used for their manufacture and the geopolitical reasons for the large increase in the prices of raw materials—namely nickel, manganese, cobalt, and aluminum.

LESMSs with batteries based on iron and other phosphates (LFP type) have a great capacity for improvement due to their environmental friendliness because they are not made with poisonous materials that are dangerous to human health. The raw materials used for their manufacture are widely available, and due to their mass production for solar PV systems, their prices will continue to fall [37]. Efforts to improve systems with LFP cells should be primarily focused on increasing their energy density and energy efficiency through the development of innovative technologies and new designs.

The future of battery technology for electric and hybrid vehicles is aimed at achieving higher energy density, accelerated charging, extended life, and improved safety and sustainability. This is achieved by using new materials for their production; developing ultra-fast charging technologies that can charge batteries up to 80% in 10–15 min; using nanomaterials to improve structural stability and extend battery life; using smart BMSs with the implementation of machine learning and artificial intelligence algorithms for charge and discharge optimization, condition monitoring, and problem prevention; using decentralized BMSs containing control systems that can function autonomously for each individual battery module; carrying out efficient recycling with the application of new methods for the extraction and reuse of valuable metals, such as lithium, cobalt, and nickel, and using environmentally friendly and sustainable materials in the production of batteries; and integration with energy storage systems, including using spent batteries from EVs and HEVs for stationary energy storage systems.

With these improvements, electric and hybrid cars will become increasingly efficient, affordable, and environmentally friendly, contributing to the faster adoption of these technologies and reducing the dependence on fossil fuels.

Author Contributions: B.V., B.D., V.D. and N.H. were involved in the full process of producing this paper, including conceptualization, methodology, modeling, validation, visualization, and preparing the manuscript. All authors have read and agreed to the published version of the manuscript.

Funding: This study was financed by the European Union—NextGenerationEU, through the National Recovery and Resilience Plan of the Republic of Bulgaria, project № BG-RRP-2.004-0005.

Data Availability Statement: Data are contained within the article.

Acknowledgments: This research was supported by the European Union—NextGenerationEU through the National Recovery and Resilience Plan of the Republic of Bulgaria, project № BG-RRP-2.004-0005.

Conflicts of Interest: The authors declare no conflicts of interest.

Abbreviations

The following abbreviations are used in this manuscript:

BMS	Battery management system
C	Capacity of cells
CC method	Coulomb counting method
ECU	Electronic control unit
EV/HEV	Electric vehicle/hybrid electric vehicle
GUI	Graphical user interface
LFP	Lithium–iron–phosphate
LESMS	Lithium-ion energy storage and management system
NCA	Lithium–cobalt–nickel–aluminum oxide
NMC	Nickel–manganese–cobalt
NCA	Nickel–cobalt–aluminum
OCV	Open-circuit voltage
SOC	State of charge
SOH	State of health

References

1. Saldaña, G.; Martín, J.I.S.; Zamora, I.; Asensio, F.J.; Oñederra, O. Analysis of the Current Electric Battery Models for Electric Vehicle Simulation. *Energies* **2019**, *12*, 2750. [CrossRef]
2. Ribeiro, P.J.G.; Dias, G.; Mendes, J.F.G. Public Transport Decarbonization: An Exploratory Approach to Bus Electrification. *World Electr. Veh. J.* **2024**, *15*, 81. [CrossRef]
3. Poggio, A.E.; Balest, J.; Zubaryeva, A.; Sparber, W. Monitored Data and Social Perceptions Analysis of Battery Electric and Hydrogen Fuelled Buses in Urban and Suburban Areas. *J. Energy Storage* **2023**, *72*, 108411. [CrossRef]
4. Velev, B. Comparative analysis of lithium-ion batteries for EV/HEV applications. *Ind. 4.0* **2018**, *3*, 73–76.
5. Blomgren, G.E. The Development and Future of Lithium Ion Batteries. *J. Electrochem. Soc.* **2016**, *164*, A5019–A5025. [CrossRef]
6. Rangarajan, S.S.; Sunddararaj, S.P.; Sudhakar, A.; Shiva, C.K.; Subramaniam, U.; Collins, E.R.; Senjyu, T. Lithium-Ion Batteries—The Crux of Electric Vehicles with Opportunities and Challenges. *Clean Technol.* **2022**, *4*, 908–930. [CrossRef]
7. Lu, L.; Han, X.; Li, J.; Hua, J.; Ouyang, M. A review on the key issues for lithium-ion battery management in electric vehicles. *J. Power Sources* **2013**, *226*, 272–288. [CrossRef]
8. Elmahallawy, M.; Elfouly, T.; Alouani, A.; Massoud, A.M. A Comprehensive Review of Lithium-Ion Batteries Modeling, and State of Health and Remaining Useful Lifetime Prediction. *IEEE Access* **2022**, *10*, 119040–119070. [CrossRef]
9. Yang, X.; Adair, K.R.; Gao, X.; Sun, X. Recent advances and perspectives on thin electrolytes for high-energy-density solid-state lithium batteries. *Energy Environ. Sci.* **2020**, *14*, 643–671. [CrossRef]
10. Aichberger, C.; Jungmeier, G. Environmental Life Cycle Impacts of Automotive Batteries Based on a Literature Review. *Energies* **2020**, *13*, 6345. [CrossRef]
11. Preger, Y.; Barkholtz, H.M.; Fresquez, A.; Campbell, D.L.; Juba, B.W.; Romàn-Kustas, J.; Ferreira, S.R.; Chalamala, B.R. Degradation of Commercial Lithium-Ion Cells as a Function of Chemistry and Cycling Conditions. *J. Electrochem. Soc.* **2020**, *167*, 120532. [CrossRef]
12. Tran, M.-K.; DaCosta, A.; Mevawalla, A.; Panchal, S.; Fowler, M. Comparative Study of Equivalent Circuit Models Performance in Four Common Lithium-Ion Batteries: LFP, NMC, LMO, NCA. *Batteries* **2021**, *7*, 51. [CrossRef]
13. Reddy, M.V.; Mauger, A.; Julien, C.M.; Paolella, A.; Zaghib, K. Brief History of Early Lithium-Battery Development. *Materials* **2020**, *13*, 1884. [CrossRef] [PubMed]
14. Fu, S.; Liu, W.; Luo, W.; Zhang, Z.; Zhang, M.; Wu, L.; Luo, C.; Lv, T.; Xie, J. State of charge estimation of lithium-ion phosphate battery based on weighted multi-innovation cubature Kalman filter. *J. Energy Storage* **2022**, *50*, 104175. [CrossRef]
15. Plett, G. Extended kalman filtering for battery management systems of LiPB-based HEV battery packs part 2 State and parameter estimation. *J. Power Sources* **2004**, *134*, 277–292. [CrossRef]
16. Zhang, Y.; Liu, Z.; Chen, Z. Smart-Leader-Based Distributed Charging Control of Battery Energy Storage Systems Considering SoC Balance. *Batteries* **2023**, *9*, 18. [CrossRef]
17. Hassan, M.; Jawad, M.; Saleem, N.; Raza, A.; Zaidi, K.; Rafiq, N. Solar Power Assisted Passive and Active Cell Balancing System: A Comprehensive Analysis. In Proceedings of the 2021 International Conference on Frontiers of Information Technology, FIT, Islamabad, Pakistan, 13–14 December 2021; pp. 224–229. [CrossRef]
18. LFP Battery in Your Next EV? Tesla, Ford Rivian Say Yes. (n.d.). Available online: https://www.recurrentauto.com/research/lfp-battery-in-your-next-ev-tesla-and-others-say-yes (accessed on 13 April 2024).
19. Premchand, M.; Gudey, S.K. Electric vehicle operation modes with reactive power support using SMC in distribution generation. *J. Energy Syst.* **2020**, *4*, 96–120. [CrossRef]
20. Sepúlveda, F.J.; Montero, I.; Barrena, F.; Miranda, M.T.; Arranz, J.I. Design and Energy Analysis of Photovoltaic-Battery Prototype Considering Different Voltage Levels. *Batteries* **2023**, *9*, 16. [CrossRef]

21. Velev, B.G.; Ivanov, I.S.; Kamenov, V.V. Experimental Study and Thermal Analysis of Cooling Systems for Brushless Motors with Double Stator and Axial Gap. *SJEE* **2021**, *18*, 333–349. [CrossRef]
22. Geisbauer, C.; Woehrl, K.; Mittmann, C.; Schweiger, H.-G. Review-Review of Safety Aspects of Calendar Aged Lithium Ion Batteries. *J. Electrochem. Soc.* **2020**, *167*, 090523. [CrossRef]
23. Whitlock, R. Lithium-Ion Battery Pack Prices Increase due to Rising Costs of Materials and Components. Available online: https://www.renewableenergymagazine.com/storage/lithiumion-battery-pack-prices-increase-due-to-20221206 (accessed on 6 December 2022).
24. Samsung INR18650-30Q 3000 mAh 20 A. (n.d.). Available online: https://www.dcpower.eu/accu-rechargeable/lithium-ion-batteries/accu-industrial/samsung-sdi/samsung-inr18650-30q (accessed on 13 April 2024).
25. Available online: http://www.batteryexpert.org/produkt/samsung-inr18650-35e-3500mah-8a/ (accessed on 10 March 2024).
26. Nickel Fuse 2P-6P Wide Continuous Roll by the Foot! 18650 Cell Level F—Battery Hookup. (n.d.). Available online: https://batteryhookup.com/products/nickel-fuse-2p-wide-continuous-roll-by-the-foot-18650-cell-level-fusing (accessed on 13 April 2024).
27. Xing, Y.; Ma, E.W.M.; Tsui, K.L.; Pecht, M. Battery Management Systems in Electric and Hybrid Vehicles. *Energies* **2011**, *4*, 1840–1857. [CrossRef]
28. Wang, M.; Le, A.V.; Noelle, D.J.; Shi, Y.; Yoon, H.; Zhang, M.; Meng, Y.S.; Qiao, Y. Effects of electrode pattern on thermal runaway of lithium-ion battery. *Int. J. Damage Mech.* **2018**, *27*, 74–81. [CrossRef]
29. Winston WB-LYP100AHA—3.3 V/100 Ah Tall LiFeYPO4 Lithium—LiFeYPO4. (n.d.). Available online: https://masori.de/en/products/wb-lyp100aha-3-3v-100ah-tall-lifeypo4 (accessed on 13 April 2024).
30. Hibms Smart Bms Active Balancer | Smart Bms 4s Lifepo4 Bluetooth—Smart Bms Balancer—Aliexpress. (n.d.). Available online: https://www.aliexpress.com/i/1005004603727710.html (accessed on 13 April 2024).
31. Zhu, W. A Smart Battery Management System for Large Format Lithium Ion Cells. Ph.D. Thesis, University of Toledo, Toledo, OH, USA, 2011.
32. Babu, P.S.; Ilango, K. Comparative Analysis of Passive and Active Cell Balancing of Li Ion Batteries. In Proceedings of the 2022 3rd International Conference on Intelligent Computing, Instrumentation and Control Technologies: Computational Intelligence for Smart Systems, ICICICT 2022, Kannur, India, 11–12 August 2022; pp. 711–716. [CrossRef]
33. Ng, K.S.; Moo, C.-S.; Chen, Y.-P.; Hsieh, Y.-C. Enhanced coulomb counting method for estimating state-of-charge and state-of-health of lithium-ion batteries. *Appl. Energy* **2009**, *86*, 1506–1511. [CrossRef]
34. Çarkit, T.; Alçi, M. Comparison of the performances of heuristic optimization algorithms PSO, ABC and GA for parameter estimation in the discharge processes of Li-NMC battery. *J. Energy Syst.* **2022**, *6*, 387–400. [CrossRef]
35. Docimo, D.J. Estimation and balancing of multi-state differences between lithium-ion cells within a battery pack. *J. Energy Storage* **2022**, *50*, 104264. [CrossRef]
36. Aimo, C.; Schmidhalter, I.; Aguirre, P. Multi-objective optimization of a lithium cell design operating in a cyclic process. *J. Energy Storage* **2023**, *57*, 106256. [CrossRef]
37. Wentker, M.; Greenwood, M.; Leker, J. A Bottom-Up Approach to Lithium-Ion Battery Cost Modeling with a Focus on Cathode Active Materials. *Energies* **2019**, *12*, 504. [CrossRef]

batteries

MDPI

Article

Using Reinforcement Learning in a Dynamic Team Orienteering Problem with Electric Batteries

Majsa Ammouriova [1,2], Antoni Guerrero [3,4], Veronika Tsertsvadze [3], Christin Schumacher [5] and Angel A. Juan [3,6,*]

1 Industrial Engineering Department, German Jordanian University, Amman 11180, Jordan; maysa.ammouri@gju.edu.jo
2 Computer Science Department, Universitat Oberta de Catalunya, 156 Rambla Poblenou, 08018 Barcelona, Spain
3 Research Center on Production Management and Engineering CIGIP, Universitat Politècnica de València, Plz. Ferrandiz-Salvador, 03801 Alcoy, Spain; antoni.guerrero@baobabsoluciones.es (A.G.); vtserts@upv.es (V.T.)
4 Baobab Soluciones, 55 Jose Abascal, 28003 Madrid, Spain
5 The Department of Business and Economics, TU Dortmund University, 44221 Dortmund, Germany; christin.schumacher@tu-dortmund.de
6 Department of Business Analytics, Euncet Business School, 1 Cami Mas Rubial, 08225 Terrassa, Spain
* Correspondence: ajuanp@upv.es

Abstract: This paper addresses the team orienteering problem (TOP) with vehicles equipped with electric batteries under dynamic travel conditions influenced by weather and traffic, which impact travel times between nodes and hence might have a critical effect on the battery capacity to cover the planned route. The study incorporates a novel approach for solving the dynamic TOP, comparing two solution methodologies: a merging heuristic and a reinforcement learning (RL) algorithm. The heuristic combines routes using calculated savings and a biased-randomized strategy, while the RL model leverages a transformer-based encoder–decoder architecture to sequentially construct solutions. We perform computational experiments on 50 problem instances, each subjected to 200 dynamic conditions, for a total of 10,000 problems solved. The results demonstrate that while the deterministic heuristic provides an upper bound for rewards, the RL model consistently yields robust solutions with lower variability under dynamic conditions. However, the dynamic heuristic, with a 20 s time limit for solving each instance, outperformed the RL model by 3.35% on average. The study highlights the trade-offs between solution quality, computational resources, and time when dealing with dynamic environments in the TOP.

Keywords: team orienteering problem; battery management; electric vehicle; reinforcement learning

1. Introduction

With the rise of electric vehicles (EVs) and unmanned aerial vehicles (UAVs), vehicle routing challenges have gained new relevance in modern disaster, security, and last-mile logistic planning [1,2]. The growing popularity of EVs has initiated a transformative shift in the automotive, vehicle, and logistics industry, providing a sustainable alternative to traditional internal combustion engine vehicles. EVs have the potential to reduce transportation costs and minimize environmental impact. In logistics, EVs must navigate complex delivery routes while managing their limited battery life and the necessity for frequent recharging, which introduces an additional layer of complexity to routing, visiting points, and delivering goods. Including battery constraints and charging cycles in modeling is essential for the practical deployment of these vehicles, whether they are delivering goods in urban areas, conducting aerial surveillance, or responding to emergencies [3]. For instance, in urban logistics, the challenge is optimizing the delivery schedule and ensuring that vehicles can complete their routes without interruptions. This problem becomes even more complex when multiple vehicles have different energy requirements [4].

In these scenarios, it is critical to optimize the coverage of regions within a limited time frame [5]. The primary challenge lies in determining optimal paths that prioritize the most relevant areas while considering the energy constraints of the vehicles. This becomes particularly complex when the power infrastructure is compromised and charging opportunities are limited [6]. These challenges can be modeled as team orienteering problems (TOPs), where the EV fleet needs to deliver goods to specific points in the area, starting at the depot. The TOP is a well-established combinatorial optimization problem, extending the classical orienteering problem [7] by introducing multiple agents or vehicles. Each team member attempts to visit candidate nodes within a prescribed time limit. Visiting a node for the first time allows the collection of a reward, and the goal is to maximize the overall team score. The problem is NP-hard even in its single-agent form, and it becomes even more difficult to solve when multiple agents are involved [8]. In real-life scenarios, these vehicles face several challenges due to dynamic conditions, such as those associated with weather status, traffic status, and travel times [9].

In scenarios where real-time routing plans are required, the ability to continuously re-optimize routes as new data become available is key. For example, if a traffic accident occurs or extreme weather conditions arise, the planned routes for EVs or UAVs may need to be adjusted on the fly. The TOP, combined with agile optimization techniques, provides a robust framework for handling these dynamic conditions, ensuring that the vehicles can continue to operate efficiently despite the changing environment [10]. By taking advantage of the power of parallel computing and biased randomization techniques, agile optimization algorithms can concisely explore a vast solution space, identifying routes that meet immediate operational needs and optimize energy usage and overall efficiency. This paper explores a dynamic TOP where vehicles must travel from an origin to a destination to maximize the number of nodes visited within the constraints of battery life. The objective is to optimize the route planning such that vehicles can collect as much reward as possible before their batteries are depleted. Figure 1 shows a dynamic scenario where an EV must travel from an origin to a destination, visiting as many nodes as possible within battery life constraints. The green circles represent visited nodes, while the pink circles are non-visited nodes. The travel time between nodes is influenced by dynamic factors such as weather conditions and road congestion, indicated by 'Tdynamic' in the figure.

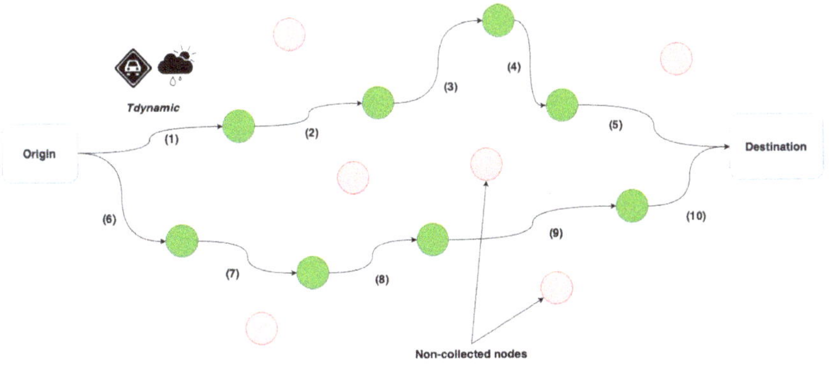

Figure 1. TOP with multiple vehicles considering battery constraints and variable travel times.

To address this dynamic version of the TOP with electric batteries, this paper proposes two approaches: one based on a biased-randomized heuristic [11] and the other using reinforcement learning (RL) [12]. The selection of reinforcement learning is primarily due to its possible suitability for dynamic problems like the TOP, where travel times fluctuate. RL allows the solving agent to adaptively learn and optimize routes under these changing conditions, potentially improving solution adaptability and quality over deterministic methods. Moreover, one of the paper's main objectives was to evaluate the performance of RL in comparison to more established approaches. The heuristic provides a solution by optimizing routes based on predefined rules, while the neural network in the RL approach is trained separately to learn and refine decisions in response to dynamic conditions. Both methods are independently tested and compared in terms of their ability to adapt to changes such as varying battery levels and travel conditions. Hence, the main contribution of this paper is the development and comparison of two approaches, heuristic-based and RL-based, for solving a TOP involving EVs with dynamic rewards. This dual approach allows us to evaluate the strengths and weaknesses of each method in making adaptive decisions in a rapidly changing environment. The remainder of this paper is structured as follows: Section 2 offers an analysis of the different EV categories, their integration into existing transportation systems, current advancements in the field, as well as factors influencing battery performance, including capacity, materials, and charging times. Section 3 summarizes the related work on algorithms designed for solving TOPs. Section 4 formally defines the specific problem. Section 5 presents the dataset used for the study. Section 6 outlines the simulation experiments and the RL algorithm applied. Section 7 details and interprets our computational findings. Finally, Section 8 concludes with a summary of the main findings and recommendations for future research.

2. Sustainable Logistics and EVs

Sustainable transportation has been included in the United Nations Sustainable Development Goal 11 (https://sdgs.un.org/topics/sustainable-transport, accessed on 29 September 2024). The transportation sector influences society development through its economic, social, and environmental impacts [13]. Additionally, it contributes heavily to gas emissions and fuel consumption. According to data collected by the European Commission (https://ec.europa.eu/eurostat/web/main/data/database, accessed on 29 September 2024), the transportation sector consumed around 31% of the total energy consumed in 2022 in the European countries, while the industry sector consumed around 25% in the same year. The transportation sector released 23% of energy-related emissions in 2019 [14]. In recent decades, sustainable transportation has become a topic for researchers investigating logistics and chain management, new fuels and vehicles, public transportation improvements, and sustainable transport alternatives [13]. These research efforts aim to optimize energy consumption in transportation and reduce gas emissions. As a result, a variety of concepts have been defined, such as ride-sharing and car-sharing, and new transportation options have been introduced to the market. EVs form one of these options aiming to reduce gas emissions as well as noise in cities [15]. In these vehicles, the conventional internal combustion engine is replaced by another one utilizing electricity from a rechargeable battery to drive the vehicle [16]. EVs are pictured as innovative symbols for decarbonizing transportation with new business opportunities and models [17]. The introduction of EVs improves air quality in terms of reducing greenhouse gas emissions and decreasing operation costs [16]. Hence, a sustainable urban environment is being developed in cities.

Several EVs types are found in the market: hybrid electric vehicles (HEVs), plug-in hybrid electric vehicles (PHEVs), battery electric vehicles (BEVs), and fuel cell electric vehicles (FCEVs). Multiple energy generation and conversion technologies are utilized in HEVs [18]. HEVs use an internal combustion engine or an electrical motor to drive the vehicle depending on the driving condition regulated by a control system. The coupling configuration of the internal combustion engine and the electric motor could be series, parallel, or series–parallel, as illustrated by Cao et al. [19]. Batteries in these vehicles are utilized to store electricity used to drive the electric motors. Real-time optimization methods are utilized to control the energy flow, including charging and discharging of the batteries. The greatest challenge in this type of EVs is the management of multiple energy sources [18]. Another type of hybrid electric vehicle is the PHEV. Similar to HEVs, PHEVs combine internal combustion engine and an electrical motor. Batteries in PHEVs might not only be charged during the drive, but they can be charged directly from the grid network [20]. BEVs utilize only electricity to drive the vehicle motor (electrical motor) [21], and the battery increases both vehicle weight and price. The driving range of BEVs depends on their batteries, which forms one of the main challenges compared to other EVs [18]. In contrast to BEVs, FCEVs have powerful input and long driving range [22]. FCEVs are EVs fueled by hydrogen that is used to generate the electricity [23]. Thus, FCEVs are supplied with hydrogen fuel tank that forms one of the challenges associated with this vehicle type (safety concerns). According to Pramuanjaroenkij and Kakaç [23], fuel cell technology is the future of transportation, with many ongoing studies.

Batteries in EVs vary in their technology, resulting in different charging speeds, driving range, and safety issues. The first kind of batteries are lead–acid batteries [16,18], consisting of a lead electrode and electrolyte. They have low gravimetric energy density (30–50 Wh/kg) and can be used for small vehicles [18]. Later, nickel-based batteries were invented [16,18], with a higher gravimetric energy density, e.g., 60–80 Wh/kg for a nickel–cadmium battery and around 140 Wh/kg for a nickel–hydrogen battery. In addition, this type of battery has a long life cycle and can be recharged many times. Many HEVs are supplied with nickel-based batteries. Lithium-ion batteries were invented in the second half of the last century [18] and are used in energy storage systems. They are characterized by a high gravimetric energy density (118–250 Wh/kg) and longer battery life. Different lithium-ion batteries types are found in the market [24]. Additionally, these batteries types varies in their cost and charging time. Various safety design guidelines for high-energy-density batteries have been explored [25]. Advances in battery chemistry, recycling, charging infrastructure, and cost reduction are continuous. The research concerning battery chemistry reflects the invention of various battery types. Since some types of EVs are being charged with electricity, charging infrastructure have been studied [16]. The location of charging outlets has become an issue for drivers using EVs. Related to this issue, recharging speed is another ambiguous aspect. The charging speed depends on the battery type, e.g., lithium-ion batteries are recharged relatively quickly. Another approach is replacing batteries in station instead of recharging them [26]. This approach demands that the issues presented by battery interchangeability, brand compatibility, and battery ownership be resolved, which will not be an easy task. Despite these challenges, the exchange of batteries provides a solution for long charging times and the need to upgrade household installations to adapt to fast charging. EVs are characterized by their driving range, which depends upon driving behavior, battery type [18], as well as the surrounding environment defined by the weather conditions, infrastructure, and traffic [27,28]. In this context, the state of charge indicates the maximum driving range according to the charge level of the batteries.

The use of EVs raises significant issues related to power distribution grid, environment, and safety. In the pursuit of sustainable development, optimization of battery operation and safety were investigated [29]. The charging of batteries in PHEVs and BEVs causes increased power demand and voltage drops [30]. Innovative approaches target supporting the power distribution grid, such as vehicle-to-grid technology. In these approaches, EVs discharge power saved in their batteries to the grid [31]. Concerning the environment, batteries consist of toxic materials and require proper disposal procedures. Additionally, thermal runaways are a concern that might lead to fires or explosions. Heating and cooling techniques for batteries were also studied [24]. Thermal management systems for batteries are crucial for controlling heating and cooling performance and the stability of batteries.

3. Related Work on Team Orienteering Problems

Traditional approaches to solve TOPs have primarily focused on deterministic and stochastic versions that assume that the environment is either entirely predictable or subject to random variations. However, real-world applications often involve dynamic changes, such as fluctuating weather conditions, varying traffic congestion, and evolving battery status, particularly relevant in the context of EVs [32]. These factors complicate the decision-making process, making it necessary to incorporate real-time data and adaptive algorithms into the solution methodology [33]. Recent research has explored various methodologies to optimize vehicle routing in dynamic and stochastic environments. For instance, recent studies have demonstrated the effectiveness of RL in navigating dynamic orienteering challenges, showing that RL-based approaches can outperform traditional heuristic methods in uncertain scenarios [34,35]. To address the complexity of dynamic TOPs, Panadero et al. [36] have investigated the use of RL combined with simheuristics to tackle dynamic and stochastic TOP scenarios, where variables such as battery life and travel times are unpredictable. Other studies have explored the use of deep reinforcement learning (DRL) for multi-vehicle TOP scenarios, demonstrating how DRL can adapt to real-time changes in operational environments [37]. Additionally, some studies have applied RL in a multi-stage TOP, emphasizing scenarios where both rewards and constraints evolve over time and offering insights into long-term planning and decision-making under uncertainty [38].

The dynamic TOP becomes particularly challenging when applied to EVs, which face constraints such as battery life and charging station availability. Effective routing in this context must take these dynamic factors into account to optimize both the travel route and the battery management system (BMS). Studies have highlighted the importance of advanced BMS to improve EV performance and longevity [39,40]. Further research has demonstrated the application of RL in this area, showing how RL can optimize load node selection and route planning under dynamic conditions [41]. Despite significant advances, challenges remain in fully addressing the dynamic TOP using RL. The high computational complexity of these problems, together with the difficulty of obtaining high-quality solutions in short computing times—especially under dynamic and stochastic conditions—continues to drive research toward more efficient algorithms [42].

4. Modeling the TOP with Dynamic Travel Times

The TOP is a classical NP-hard problem where a set of nodes and vehicles are given, and the main goal is to maximize the sum of the reward collected from visiting each node. A maximum travel time is allowed for each vehicle, which means that not all nodes can be visited. This travel time can be limited by various factors, such as the battery capacity of EVs. Moreover, each vehicle has to exit from the initial depot and end its route at the final one. In a formal way, let $G = (N, E)$ be a directed graph, where $N = \{1, 2, \ldots, n\} \cup \{o, d\}$ represents the set of nodes, with o and d denoting the origin and destination depots, respectively. The set of directed edges is given by $E = \{(i, j) \mid i, j \in N\}$, which represents the possible paths between nodes. Each node $i \in N$ has an associated reward r_i for visiting that node, with the rewards at the origin and destination depots, r_o and r_d, set to zero. Let V be the set of all vehicles. A binary decision variable, x_{ijv}, takes the value of 1 if vehicle v travels from node i to node j and 0 otherwise. Then, the mathematical model for TOP with dynamic travel times is formulated as follows:

$$\max \sum_{v \in V} \sum_{i,j \in N} x_{ijv} r_j \tag{1}$$

Equation (1) defines the objective function, which aims to maximize the total reward collected from the visited nodes. The following constraints apply:

$$\sum_{j \in N} x_{ojv} \leq 1 \qquad \forall v \in V \tag{2}$$

$$\sum_{i \in N} x_{idv} = \sum_{j \in N} x_{ojv} \qquad \forall v \in V \tag{3}$$

$$x_{ijv} \leq \sum_{j \in N} x_{ojv} \qquad \forall v \in V \tag{4}$$

$$\sum_{i \in N} x_{iov} + \sum_{j \in N} x_{djv} = 0 \qquad \forall v \in V \tag{5}$$

Equation (2) ensures that a vehicle departs the origin at most once. Equations (2) and (3) state that if a vehicle departs from the origin depot, it must eventually arrive at the destination depot. Meanwhile, Equation (4) ensures that a vehicle can visit other nodes only if it leaves the origin depot, and Equation (5) enforces that no vehicle can revisit the origin depot or depart from the destination depot, ensuring that the route starts at the origin and ends at the destination. Additional constraints are described next:

$$\sum_{v \in V} \sum_{i \in N} x_{ijv} \leq 1 \qquad \forall j \in N \setminus \{d\} \tag{6}$$

$$\sum_{i \in N} x_{ijv} = \sum_{i \in N} x_{jiv} \qquad \forall j \in N \setminus \{o, d\}, v \in V \tag{7}$$

$$y_{iv} - y_{jv} + 1 \leq (1 - x_{ijv})|N| \qquad \forall i, j \in N, v \in V \tag{8}$$

Equation (6) ensures that each node is visited at most once, while Equation (7) states that if a vehicle arrives at a node, it must also depart from that node. Moreover, in Equation (8) the variables y_{iv} are introduced, which represent the order of the node i in the route of vehicle v. This restriction ensures that there are no subtours. A last set of equations is introduced next:

$$\sum_{i,j \in N} x_{ijv} f(i,j) \leq L \qquad \forall v \in V \tag{9}$$

$$y_{iv} \geq 0, \qquad \forall i \in N, v \in V \tag{10}$$

$$x_{ijv} \in \{0, 1\} \qquad \forall i, j \in N, v \in V \tag{11}$$

Equation (9) ensures that the travel time for each vehicle does not exceed the maximum allowable travel time L. Here, $f(i,j)$ represents the travel time between nodes i and j, which is dynamic and varies according to the specific conditions of the edge. Equations (10) and (11) enforce the nature of the variables.

5. A Numerical Case Study

Consider a TOP with dynamic travel times, which are influenced by weather and traffic conditions. The positions of the nodes are known, as are the deterministic travel times. Since this is a case study, we are supposing that these travel times can increase by up to 12.5%, depending on the dynamic conditions between each pair of nodes. For illustrative purposes, let us assume that the travel time can be modeled using a linear regression. In practical applications, actual travel data could indeed be leveraged to provide more precise predictions. However, implementing such a model was beyond the scope of this study, which focused on evaluating the two different methods: heuristics and RL models. The weather and traffic values range from 0 to 1, depending on their influence on the travel time increase. Hence, we will assume that the 'real' travel time between nodes i and j is given by $f(i,j)$:

$$f(i,j) = t_{ij}(1 + w_{ij} \cdot 0.0625 + h_{ij} \cdot 0.0625)$$

In this case, t_{ij} represents the deterministic travel time between nodes i and j, w_{ij} represents the weather conditions, and h_{ij} represents the traffic conditions between the pair of nodes. A value of 0 indicates favorable weather or traffic conditions, while 1 represents the worst conditions. It is assumed that weather and traffic conditions contribute equally to the travel time increase (each with the coefficient 0.0625). Each node has a reward that also varies between 0 and 1, and the x and y coordinates of each node are randomly generated within the range [0, 1]. Additionally, the maximum travel time allowed for each vehicle is randomly generated, but the minimum allowed travel time is set to be 1.125 times the travel time between the two depots. This function, f, is used solely for calculating travel times and can be more complex to better reflect real-world conditions. However, the exact nature of this function remains unknown to the solving algorithm. In this study, 50 different problem instances have been defined. These problem instances differ in node location

(x and y coordinates), vehicle maximum travel time, and node rewards. A total of 20 nodes are found in each instance, and these nodes can be visited by one of two vehicles to collect rewards from them. Likewise, the dynamic conditions (weather and traffic) have been varied 200 times for each instance. It is important to recall that each problem instance is defined by a specific set of nodes, rewards, and depot locations, and then it is solved 200 times, each one under different dynamic conditions (which means different travel times between node pairs due to changing factors). The reported reward for each instance represents the average reward collected across these 200 different dynamic condition.

6. Solving Approaches

The heuristic approach is inspired by the method proposed by Panadero et al. [8]. It is a merging heuristic, where an initial dummy solution is created, consisting of simple routes from the origin depot to a node and then to the destination depot, and routes are then combined based on calculated savings. The savings $s_{i,j}$ for each pair of nodes i and j are computed as follows:

$$s_{i,j} = \alpha(c_{0,j} + c_{i,n+1} - c_{i,j}) + (1-\alpha)(r_i + r_j) \quad (12)$$

In this equation, $c_{0,j}$ represents the travel cost from the initial depot to node j, $c_{i,n+1}$ represents the cost of traveling from node i to the destination depot, and $c_{i,j}$ represents the travel cost between nodes i and j. Additionally, r_i and r_j denote the rewards associated with nodes i and j, respectively. The merging process incorporates a biased-randomized strategy [11], meaning that merges are not performed in a purely greedy way, but rather in a randomized manner that aligns with the heuristic's logic. Specifically, a geometric distribution is applied for the biased-randomization, with a parameter β that varies randomly between 0.1 and 0.3. This heuristic, in conjunction with the methodology shown in Figure 2, is used to solve the dynamic TOP.

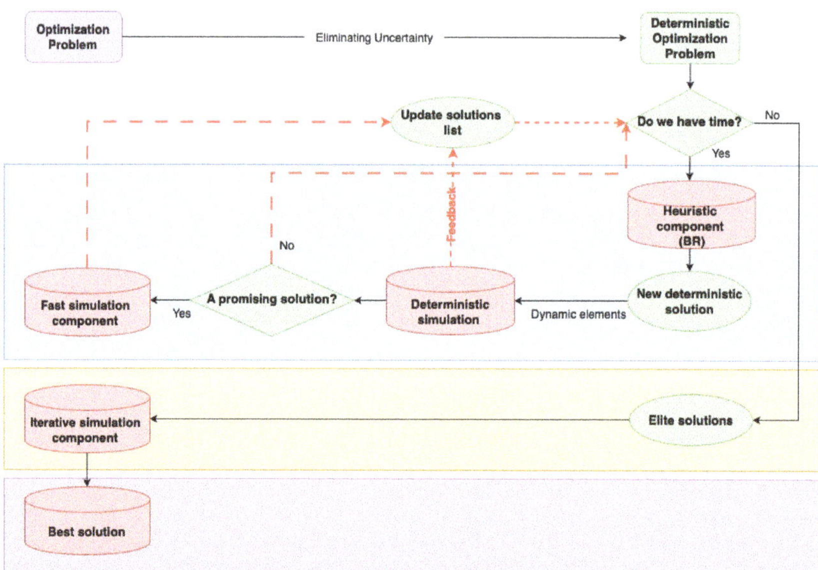

Figure 2. Schema of the heuristic utilized to solve the TOP.

Since the heuristic is randomized, a maximum time limit is set for solving each instance. The routes are merged based on their savings; instead of always choosing the best merge, it is performed in a randomized way. The merges are stored in a list and sorted based on their savings, and a probability is assigned to each one of them following a geometric distribution with parameter β. This parameter varies between 0.1 and 0.3, as we found it provides the best solutions. Then, the routes are chosen based on that probability and merged. This allows the algorithm to find better solutions as it tries to escape local optima. With sufficient remaining time, the heuristic addresses the deterministic problem in a biased-randomized manner. If the resulting solution is promising, it is deterministically simulated several times across various dynamic environments. This component simulates dynamic elements such as traffic and weather conditions, which impact travel times, to test the robustness of the solution given by the heuristic. If the solution becomes infeasible under the dynamic conditions, its reward is set to zero. Otherwise, the solution and its associated mean reward are stored in a solution list. If the result is not promising or the short deterministic simulation is finished, the process continues iteratively until the maximum allowed time is reached. Once it is reached, the top 10 solutions from the solution list are tested in dynamic environments with a higher number of deterministic simulations. The final solution is selected from this elite group based on which performs best during these extensive tests.

The RL method, however, takes a different approach. It is built on the transformer architecture introduced by Vaswani [43], which has recently gained significant attention for solving NP-hard problems. The method follows an encoder-decoder structure, where the encoder processes variable-length input data and the decoder generates the solution. Regarding the choice of a transformer-based encoder-decoder architecture, this architecture was chosen for its ability to process input data with variable sequence lengths, making it particularly well-suited to problems like the TOP, where each problem instance can contain a different number of nodes. For a more detailed explanation, refer to Berto et al. [44], which presents state-of-the-art benchmarks, ideas, and techniques for modeling NP-hard problems such as the TOP using RL. In this case, the RL algorithm follows a constructive approach, and the flowchart of its process is shown in Figure 3.

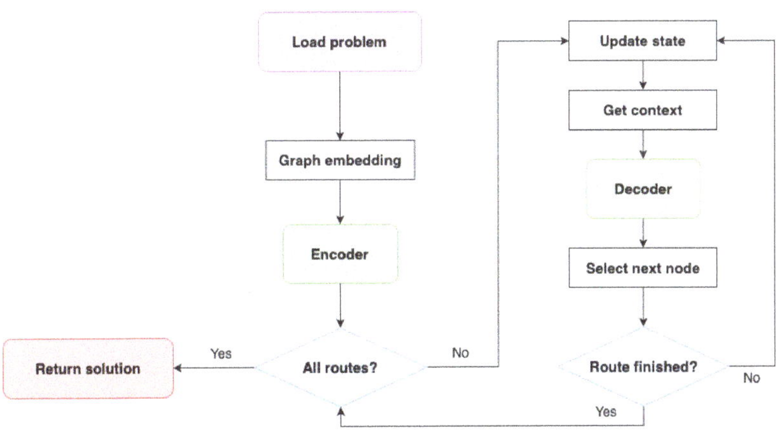

Figure 3. Schema of the RL algorithm.

First, the problem data (the position and reward of each node) are loaded into the algorithm. These data are then embedded using a linear network, with the resulting outputs passed to the encoder, which follows the architecture of the transformer encoder. The solution is constructed sequentially, route by route, and node by node. To determine the next node to visit, the algorithm updates the current state of the problem—such as the vehicle's location, the nodes available for visitation, and the current battery level. This updated state, which includes information about the graph, the vehicle's current position, the nodes already visited, and the remaining battery capacity, serves as the context provided to the model. Based on this context, the model decides which node to visit next. This process continues until the vehicle reaches the destination depot, at which point the algorithm moves on to the next route, repeating the process until all routes are completed.

For training this algorithm, the reward maximization serves as the objective function, guiding the agent's learning process to prioritize routes that provide the highest possible reward while respecting the maximum travel time. Notice that the solution generated by the RL algorithm is represented as a permutation $\pi = (\pi_1, \pi_2, \ldots)$ of a subset of nodes, where the final depot can be visited more than once, but not all nodes are required to be included in the route. Policy-gradient methods learn the policy through gradient-based optimization techniques. In this framework, a stochastic policy $p(\pi|s)$ is defined to select a solution π based on a given problem s. This policy is factorized and parameterized by θ as follows:

$$p_\theta(\pi|s) = \prod_{t=1}^{N} p_\theta(\pi_t|s, \pi_{1:t-1}) \tag{13}$$

For training the model's policy, Williams [45] introduced a policy gradient estimator using Monte Carlo sampling, under the assumption that rewards are independent of θ. This approach is known as the REINFORCE algorithm:

$$\nabla_\theta \mathcal{L}(\theta|s) = \mathbb{E}_{p_\theta(\pi|s)}[R(\pi)\nabla_\theta \log p_\theta(\pi|s)] \tag{14}$$

However, a significant drawback of this method is its high variance, which can make model training less efficient and more unstable. To mitigate this issue, a baseline value $b(s)$ is incorporated into Equation (14), reducing variance and thereby improving training stability and overall performance:

$$\nabla_\theta \mathcal{L}(\theta|s) = \mathbb{E}_{p_\theta(\pi|s)}[(R(\pi) - b(s))\nabla_\theta \log p_\theta(\pi|s)] \tag{15}$$

The baseline used in our case is the one proposed by Lee and Ahn [38], where problem instances and their equivalent variations, generated through augmentations, are utilized to improve the model's ability to generalize more effectively. The training phase is divided into epochs, with each epoch consisting of 2000 training steps and a batch size of 256. At the end of each epoch, the model is evaluated on 100,000 randomly generated instances to assess its performance, and if the mean reward from the validation set exceeds that of the current best model, a t-test with $\alpha = 0.05$ is performed to confirm if the new model is statistically superior. If the new model is found to be better, it is saved and used as the new baseline. The goal of the training process is for the model to continually compete against its best previous version. For the training hyperparameters, a learning rate of 5×10^{-5} was used, along with the Adam optimizer [46]. Although we experimented with higher learning rate values, the model either failed to converge or did not behave as expected during training. The training was executed on a workstation equipped with 32 GB of RAM and an NVIDIA 4060 GPU.

7. Computational Experiments

This section illustrates the experiment results after testing the algorithm described in Section 6. According to Section 5, 50 problem instances are solved, and 200 different dynamic conditions form problems for each instance, resulting in total 10,000 problems solved. The heuristic used to solve the TOP was allocated 2 s and 20 s for each instance to achieve the best result. This means that the heuristic solves each instance during 2 and 20 s, respectively. Since the heuristic is randomized, this ensures that a broader set of solutions is obtained, of which the best is returned. Those running times were chosen to show how the solution improves when letting the heuristic run for more time; moreover, the RL method takes about 0.5 s to obtain the solution, meaning that it is possible to compare computational efficiency between different solving methods. Allowing more than 20 s for the heuristic was also tested, but it did not improve performance. Additionally, the deterministic version was also solved to evaluate the impact of incorporating dynamic conditions. This version of the heuristic also runs for 2 s before returning the solution. A reward of 0 is given when the solution under dynamic conditions becomes infeasible.

Four different experiments have been run: 'deterministic' refers to the heuristic under deterministic conditions; 'dynamic heu. (20 s)' indicates the heuristic solving the problems under dynamic conditions with a time limit of 20 s for each instance; 'dynamic heu. (2 s)' refers the heuristic under dynamic conditions and solving the problems within 2 s; and 'dynamic RL' refers to using RL to solve the problems under dynamic conditions. Notice that the deterministic heuristic approach is applied to a static version of the problem, whereas both the dynamic heuristic and the RL model solve the problem under dynamic conditions. Therefore, it is expected that the deterministic heuristic provides higher rewards, as it does not have to adjust to the uncertainties present in the dynamic scenario. We included the deterministic results intentionally to serve as a benchmark, helping to illustrate the impact that dynamic elements have on overall performance. Both the dynamic heuristic and the RL approach attempt to solve the problem without full knowledge of the dynamic conditions, which can vary significantly in real-world scenarios, especially when travel times are high enough. While better results could be achieved, this would require an estimation of the dynamic conditions beforehand.

The 50 problem instances are generated randomly: both the position of the nodes and their reward, as well as the maximum travel time allowed for each vehicle. Figure 4 shows the results obtained after running the four experiments. It is observed that the highest reward was collected by vehicles in the deterministic experiment. Both the mean and the median are the highest in this experiment compared to the other three experiments. By taking a closer look at the experiments under dynamic conditions, it is clear that the mean of the collected rewards in the dynamic RL and dynamic heu. (20 s) experiments are close and greater than those collected in dynamic heu. (2 s). The median of rewards collected in the dynamic RL experiment is the greatest. Additionally, lower variability of collected reward values is noticed in the dynamic RL experiment. This variability can be presented by the inter-quartile distance in Figure 4. The lower variability indicates greater solution reliability and ability to find consistent solutions. The greater variability in solutions found by the heuristic indicates the difficulty in finding reliable solutions.

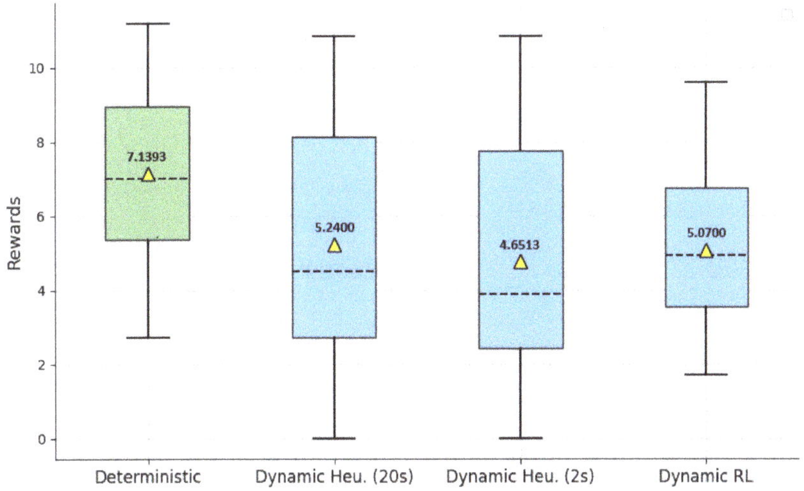

Figure 4. Distribution of results obtained in the defined four experiments.

A deeper investigation of the result is presented in Table 1. Table 1 compare between the four experiments by recording the mean of found solutions per a problem instance. The average collected reward for 21 problem instances out of the 50 instances are tabulated. These means are related to the 200 different defined dynamic condition combinations in a problem instance. Similar to the observation in Figure 4, the dynamic heu. (20 s) experiments showed clearly better performance due to the greater exploration time compared to dynamic heu. (2 s). On average, it outperformed the dynamic RL experiment by 3.35%. Those 21 problem instances presented in Table 1 were chosen randomly from the 50 instances solved. The main goal is to show how the heuristic and RL model behave in different scenarios.

Table 1. Comparison of mean results for each problem across four solution approaches: the deterministic heuristic, the dynamic heuristic using 20 s, the dynamic heuristic using 2 s, and the RL model.

Problem	Deterministic	Dynamic Heu. (20 s)	Dynamic Heu. (2 s)	Dynamic RL
0	4.2553	3.1504	2.1379	3.5322
1	8.7905	8.1820	4.6382	5.4216
2	4.0850	2.6747	1.8263	2.8274
3	6.4999	6.3844	3.2809	4.8508
4	8.3060	6.5841	6.4133	6.4047
5	5.8537	3.8344	3.7612	3.1818
6	9.0080	9.0080	9.0080	7.7950
7	11.1873	10.7949	10.7949	8.4732
8	6.6231	5.1850	5.1850	4.9351
9	4.7175	2.0463	2.0240	3.4620
10	9.2883	5.3923	0.0000	6.1189
11	6.6812	0.0694	0.1005	3.8277
12	5.3447	3.1942	3.1942	3.3079
13	8.8668	8.8668	8.8668	7.1570
14	2.7068	1.5510	1.5510	2.0699
15	5.3801	5.0151	5.0151	4.3966
16	7.0243	3.3182	3.2398	4.1747
17	4.0835	2.4231	2.4231	2.9224
18	5.8006	2.0723	2.0723	3.8718
19	7.7680	4.4660	3.0630	5.0875
20	4.7668	4.4149	3.5105	3.9444
⋮	⋮	⋮	⋮	⋮
Avg	**7.1393**	**5.2400**	**4.6513**	**5.0700**

The solution found for the deterministic version of the problem instances is considered as the upper bound. With introduced dynamic conditions, the vehicle travel time increases and reduces the possibility to collect all rewards as in the deterministic version of the problem. Depending on the severity of the weather and traffic conditions, the collected rewards are affected and reduced accordingly. In some problem instances, such as instance 6 and 13 in Table 1, the heuristic was successful in finding the greatest reward under the dynamic conditions, while the RL model was not able to find. In other instances, such as instances 1 and 9, the RL model was successful in solutions with rewards greater than those found by the heuristic under the dynamic conditions. In Figure 5, the training validation reward of the RL model is illustrated. As expected, the model shows a rapid improvement during the initial epochs. However, the rate of improvement significantly decreases in later epochs. The red dots show the epochs when the model improves, at which point the new baseline is saved.

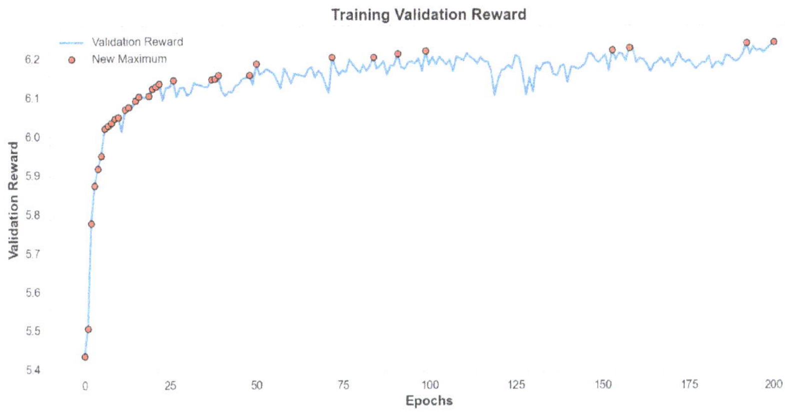

Figure 5. Validation reward at each epoch in the training phase.

It is also important to highlight the time required to train the model. In this case, training for 200 epochs took approximately 16 h. Additionally, a significant amount of training data is needed, which can be challenging to generate, particularly for more complex problems that simulate intricate real-world conditions and are hard to validate. The scalability of the RL model also requires attention, as varying the number of nodes from those used during training may result in sub-optimal performance. According to the results, the training of the RL model is challenging and demands resources. Once the RL model is trained, it provides promising and reliable solutions. The solution reliability is presented by the lower variability compared to the other approaches (Figure 4).

8. Conclusions

This paper presents a dual approach to solve the dynamic TOP with EVs, constrained by battery driving range and impacted by real-time dynamic conditions. To address this challenge, we propose two independent methodologies: a heuristic-based approach for rapid solution generation and a reinforcement learning approach for adaptive decision-making. By applying these distinct methods, we are able to explore the strengths and weaknesses of each in scenarios where dynamic factors like road congestion, battery status, and travel times fluctuate continuously.

The heuristic approach efficiently produces initial solutions, particularly in relatively stable environments, providing a fast and practical option for route planning. In contrast, the reinforcement learning approach excels in more dynamic environments, learning to adapt decisions as conditions evolve, and consistently achieving higher-quality solutions by optimizing both the number of nodes visited and energy efficiency. The computational experiments reveal that while the heuristic approach performs well under deterministic conditions, it might lack the flexibility needed to adapt in highly variable scenarios. The reinforcement learning approach, however, demonstrates its capacity to incorporate dynamic elements, offering robust and reliable solutions even under significant uncertainty. Although RL experiment showed a great performance in this study in solving the TOP, the training of the RL model is computationally demanding.

In this study, the RL approach is used to solve TOP as an example of a last-mile delivery problem, involving dynamic components. The dynamic travel time represents one of real-time aspects encountered in optimization problems in addition to stochastic uncertainty. The RL approach showed the ability to handle the dynamic conditions and recommended solutions after training the RL model to handle such problems. Similar applications could be found in other last-mile delivery problems, especially those reflecting real-time and real-world problems.

Our future work will focus on introducing uncertainty along with the dynamic conditions to the TOP problem, especially in regard to battery management and duration. Additionally, investigating approaches for speeding up training of the reinforcement learning model is also a line to consider. The scalability of the solving approach is an issue to further investigate and evaluate in future works. In real-life problems, the batteries life variability is one of aspects to be considered. Accordingly, the future work can expand the problem definition and consider batteries life as well.

Author Contributions: Conceptualization, A.A.J.; methodology, A.G., V.T. , C.S. and M.A.; software, A.G.; validation, M.A. and C.S.; writing—original draft preparation, V.T., A.G. and C.S.; writing—review and editing, V.T., M.A. and A.A.J.; supervision, V.T. and M.A. All authors have read and agreed to the published version of the manuscript.

Funding: The present work has been partially funded by the Spanish Government (IA4TES project 'Artificial Intelligence for Sustainable Energy Transition'), the Spanish Ministry of Science-AEI (PID2022-138860NB-I00 and RED2022-134703-T), as well as the European Commission (SUN HORIZON-CL4-2022-HUMAN-01-14-101092612, AIDEAS HORIZON-CL4- 2021-TWIN-TRANSITION-01-07-101057294).

Data Availability Statement: Data are contained within the article.

Conflicts of Interest: The authors declare no conflicts of interest.

Abbreviations

The following abbreviations are used in this manuscript:

BEV	battery electric vehicle
DRL	deep reinforcement learning
EV	electric vehicle
FCEV	fuel cell electric vehicles
HEV	hybrid electric vehicle
PHEV	plug-in hybrid electric vehicle
RL	reinforcement learning
UAV	unmanned aerial vehicle

References

1. Puzicha, A.; Buchholz, P. Dynamic Mission Control for Decentralized Mobile Robot Swarms. In Proceedings of the 2022 IEEE International Symposium on Safety, Security, and Rescue Robotics (SSRR), Sevilla, Spain, 8–10 November 2022; pp. 257–263.
2. Rabe, M.; Poeting, M.; Klueter, A. Evaluating the Benefits of Collaborative Distribution with Supply Chain Simulation. In *Food Supply Chains in Cities: Modern Tools for Circularity and Sustainability*; Palgrave Macmillan: Cham, Switzerland, 2020; pp. 69–100.
3. Poeting, M.; Prell, B.; Rabe, M.; Uhlig, T.; Wenzel, S. Considering energy-related factors in the simulation of logistics systems. In Proceedings of the 2019 Winter Simulation Conference (WSC), National Harbor, MD, USA, 8–11 December 2019; pp. 1849–1858.
4. Poeting, M.; Schaudt, S.; Clausen, U. A comprehensive case study in last-mile delivery concepts for parcel robots. In Proceedings of the 2019 Winter Simulation Conference (WSC), National Harbor, MD, USA, 8–11 December 2019; pp. 1779–1788.
5. Khan, A.; Zhang, J.; Ahmad, S.; Memon, S.; Qureshi, H.A.; Ishfaq, M. Dynamic positioning and energy-efficient path planning for disaster scenarios in 5G-assisted multi-UAV environments. *Electronics* **2022**, *11*, 2197. [CrossRef]
6. Khan, S.I.; Qadir, Z.; Munawar, H.S.; Nayak, S.R.; Budati, A.K.; Verma, K.D.; Prakash, D. UAVs path planning architecture for effective medical emergency response in future networks. *Phys. Commun.* **2021**, *47*, 101337. [CrossRef]
7. Golden, B.; Levy, L.; Vohra, R. The Orienteering Problem. *Nav. Res. Logist.* **1987**, *34*, 307–318. [CrossRef]
8. Panadero, J.; Currie, C.; Juan, A.A.; Bayliss, C. Maximizing Reward from a Team of Surveillance Drones under Uncertainty Conditions: A Simheuristic Approach. *Eur. J. Ind. Eng.* **2020**, *14*, 1–23. [CrossRef]
9. Sebai, M.; Rejeb, L.; Denden, M.A.; Amor, Y.; Baati, L.; Said, L.B. Optimal electric vehicles route planning with traffic flow prediction and real-time traffic incidents. *Int. J. Electr. Comput. Eng. Res.* **2022**, *2*, 1–12. [CrossRef]
10. Peng, Z.; Li, B.; Chen, X.; Wu, J. Online route planning for UAV based on model predictive control and particle swarm optimization algorithm. In Proceedings of the 10th World Congress on Intelligent Control and Automation, Beijing, China, 6–8 July 2012; pp. 397–401.
11. Juan, A.A.; Keenan, P.; Martí, R.; McGarraghy, S.; Panadero, J.; Carroll, P.; Oliva, D. A review of the role of heuristics in stochastic optimisation: From metaheuristics to learnheuristics. *Ann. Oper. Res.* **2023**, *320*, 831–861. [CrossRef]
12. Szepesvári, C. *Algorithms for Reinforcement Learning*; Springer Nature: Cham, Switzerland, 2022.

13. Zhao, X.; Ke, Y.; Zuo, J.; Xiong, W.; Wu, P. Evaluation of sustainable transport research in 2000–2019. *J. Clean. Prod.* **2020**, *256*, 120404. [CrossRef]
14. IPCC. *Climate Change 2022: Mitigation of Climate Change*; Contribution of Working Group III to the Sixth Assessment Report of the Intergovernmental Panel on Climate Change; Cambridge University Press, Cambridge, UK; New York, NY, USA, 2022.
15. IEA. *Global EV Outlook 2023*; International Energy Agency: Paris, France, 2023.
16. Alanazi, F. Electric vehicles: Benefits, challenges, and potential solutions for widespread adaptation. *Appl. Sci.* **2023**, *13*, 6016. [CrossRef]
17. Corradi, C.; Sica, E.; Morone, P. What drives electric vehicle adoption? Insights from a systematic review on European transport actors and behaviours. *Energy Res. Soc. Sci.* **2023**, *95*, 102908. [CrossRef]
18. Ntombela, M.; Musasa, K.; Moloi, K. A comprehensive review for battery electric vehicles (BEV) drive circuits technology, operations, and challenges. *World Electr. Veh. J.* **2023**, *14*, 195. [CrossRef]
19. Cao, Y.; Yao, M.; Sun, X. An overview of modelling and energy management strategies for hybrid electric vehicles. *Appl. Sci.* **2023**, *13*, 5947. [CrossRef]
20. Martinez, C.M.; Hu, X.; Cao, D.; Velenis, E.; Gao, B.; Wellers, M. Energy management in plug-in hybrid electric vehicles: Recent progress and a connected vehicles perspective. *IEEE Trans. Veh. Technol.* **2016**, *66*, 4534–4549. [CrossRef]
21. König, A.; Nicoletti, L.; Schröder, D.; Wolff, S.; Waclaw, A.; Lienkamp, M. An overview of parameter and cost for battery electric vehicles. *World Electr. Veh. J.* **2021**, *12*, 21. [CrossRef]
22. Waseem, M.; Amir, M.; Lakshmi, G.S.; Harivardhagini, S.; Ahmad, M. Fuel cell-based hybrid electric vehicles: An integrated review of current status, key challenges, recommended policies, and future prospects. *Green Energy Intell. Transp.* **2023**, *2*, 100121. [CrossRef]
23. Pramuanjaroenkij, A.; Kakaç, S. The fuel cell electric vehicles: The highlight review. *Int. J. Hydrogen Energy* **2023**, *48*, 9401–9425. [CrossRef]
24. Khan, A.; Yaqub, S.; Ali, M.; Ahmad, A.W.; Nazir, H.; Khalid, H.A.; Iqbal, N.; Said, Z.; Sopian, K. A state-of-the-art review on heating and cooling of lithium-ion batteries for electric vehicles. *J. Energy Storage* **2024**, *76*, 109852. [CrossRef]
25. Duan, J.; Tang, X.; Dai, H.; Yang, Y.; Wu, W.; Wei, X.; Huang, Y. Building safe lithium-ion batteries for electric vehicles: A review. *Electrochem. Energy Rev.* **2020**, *3*, 1–42. [CrossRef]
26. Ahmad, F.; Saad Alam, M.; Saad Alsaidan, I.; Shariff, S.M. Battery swapping station for electric vehicles: Opportunities and challenges. *IET Smart Grid* **2020**, *3*, 280–286. [CrossRef]
27. Li, W.; Stanula, P.; Egede, P.; Kara, S.; Herrmann, C. Determining the main factors influencing the energy consumption of electric vehicles in the usage phase. *Procedia CIRP* **2016**, *48*, 352–357. [CrossRef]
28. Bi, J.; Wang, Y.; Zhang, J. A data-based model for driving distance estimation of battery electric logistics vehicles. *EURASIP J. Wirel. Commun. Netw.* **2018**, *2018*, 251. [CrossRef]
29. Togun, H.; Aljibori, H.S.S.; Biswas, N.; Mohammed, H.I.; Sadeq, A.M.; Rashid, F.L.; Abdulrazzaq, T.; Zearah, S.A. A critical review on the efficient cooling strategy of batteries of electric vehicles: Advances, challenges, future perspectives. *Renew. Sustain. Energy Rev.* **2024**, *203*, 114732. [CrossRef]
30. Brenna, M.; Foiadelli, F.; Leone, C.; Longo, M. Electric Vehicles Charging Technology Review and Optimal Size Estimation. *J. Electr. Eng. Technol.* **2020**, *15*, 2539–2552. [CrossRef]
31. Yong, J.Y.; Ramachandaramurthy, V.K.; Tan, K.M.; Mithulananthan, N. A Review on the State-of-the-Art Technologies of Electric Vehicle, Its Impacts and Prospects. *Renew. Sustain. Energy Rev.* **2015**, *49*, 365–385. [CrossRef]
32. Montoya, A.; Guéret, C.; Mendoza, J.E.; Villegas, J.G. The electric vehicle routing problem with nonlinear charging function. *Transp. Res. Part Methodol.* **2017**, *103*, 87–110. [CrossRef]
33. Zografos, K.G.; Androutsopoulos, K.N.; Vasilakis, G.M. A real-time decision support system for roadway network incident response logistics. *Transp. Res. Part C Emerg. Technol.* **2002**, *10*, 1–18. [CrossRef]
34. Xu, Y.; Fang, M.; Chen, L.; Xu, G.; Du, Y.; Zhang, C. Reinforcement learning with multiple relational attention for solving vehicle routing problems. *IEEE Trans. Cybern.* **2021**, *52*, 11107–11120. [CrossRef] [PubMed]
35. Vincent, F.Y.; Salsabila, N.Y.; Lin, S.W.; Gunawan, A. Simulated annealing with reinforcement learning for the set team orienteering problem with time windows. *Expert Syst. Appl.* **2024**, *238*, 121996.
36. Panadero, J.; Juan, A.A.; Ghorbani, E.; Faulin, J.; Pagès-Bernaus, A. Solving the stochastic team orienteering problem: Comparing simheuristics with the sample average approximation method. *Int. Trans. Oper. Res.* **2024**, *31*, 3036–3060. [CrossRef]
37. Sankaran, P. *Deep Reinforcement Learning and Hybrid Approaches to Solve Multi-Vehicle Combinatorial Optimization Problems*; Rochester Institute of Technology: Rochester, NY, USA, 2023.
38. Lee, D.H.; Ahn, J. Multi-start team orienteering problem for UAS mission re-planning with data-efficient deep reinforcement learning. *Appl. Intell.* **2024**, *54*, 4467–4489. [CrossRef]
39. Wang, Y.; Zhou, J.; Sun, Y.; Fan, J.; Wang, Z.; Wang, H. Collaborative multidepot electric vehicle routing problem with time windows and shared charging stations. *Expert Syst. Appl.* **2023**, *219*, 119654. [CrossRef]
40. Sánchez, D.G.; Tabares, A.; Faria, L.T.; Rivera, J.C.; Franco, J.F. A clustering approach for the optimal siting of recharging stations in the electric vehicle routing problem with time windows. *Energies* **2022**, *15*, 2372. [CrossRef]
41. Juan, A.A.; Marugan, C.A.; Ahsini, Y.; Fornes, R.; Panadero, J.; Martin, X.A. Using Reinforcement Learning to Solve a Dynamic Orienteering Problem with Random Rewards Affected by the Battery Status. *Batteries* **2023**, *9*, 416. [CrossRef]

42. Wang, R.; Liu, W.; Li, K.; Zhang, T.; Wang, L.; Xu, X. Solving Orienteering Problems by Hybridizing Evolutionary Algorithm and Deep Reinforcement Learning. *IEEE Trans. Artif. Intell.* **2024**, *5*, 5493–5508. [CrossRef]
43. Vaswani, A.; Shazeer, N.; Parmar, N.; Uszkoreit, J.; Jones, L.; Gomez, A.N.; Kaiser, L.u.; Polosukhin, I. Attention is all you need. *Adv. Neural Inf. Process. Syst.* **2017**, *30*, 5998–6008.
44. Berto, F.; Hua, C.; Park, J.; Luttmann, L.; Ma, Y.; Bu, F.; Wang, J.; Ye, H.; Kim, M.; Choi, S.; et al. Rl4co: An extensive reinforcement learning for combinatorial optimization benchmark. *arXiv* **2023**, arXiv:2306.17100.
45. Williams, R.J. Simple statistical gradient-following algorithms for connectionist reinforcement learning. *Mach. Learn.* **1992**, *8*, 229–256. [CrossRef]
46. Ruthotto, L.; Haber, E. An introduction to deep generative modeling. *GAMM-Mitteilungen* **2021**, *44*, e202100008. [CrossRef]

batteries

MDPI

Article

Analysis of Aging and Degradation in Lithium Batteries Using Distribution of Relaxation Time

Muhammad Sohaib, Abdul Shakoor Akram and Woojin Choi *

School of Electrical Engineering, Soongsil University, Seoul 06978, Republic of Korea;
sohaib@soongsil.ac.kr (M.S.); shakoor@soongsil.ac.kr (A.S.A.)
* Correspondence: cwj777@ssu.ac.kr

Abstract: In this paper, the deconvolution of Electrochemical Impedance Spectroscopy (EIS) data into the Distribution of Relaxation Times (DRTs) is employed to provide a detailed examination of degradation mechanisms in lithium-ion batteries. Using an nth RC model with Gaussian functions, this study achieves enhanced separation of overlapping electrochemical processes where Gaussian functions yield smoother transitions and clearer peak identification than conventional piecewise linear functions. The advantages of employing Tikhonov Regularization (TR) with Gaussian functions over Maximum Entropy (ME) and FFT methods are highlighted as this approach provides superior noise resilience, unbiased analysis, and enhanced resolution of critical features. This approach is applied to LIB cell data to identify characteristic peaks of the DRT plot and evaluate their correlation with battery degradation. By observing how these peaks evolve through cycles of battery aging, insights into specific aging mechanisms and performance decline are obtained. This study combines experimental measurements with DRT peak analysis to characterize the impedance distribution within LIBs which enables accelerated detection of degradation pathways and enhances the predictive accuracy for battery life and reliability. This analysis contributes to a refined understanding of LIB degradation behavior, supporting the development of advanced battery management systems designed to improve safety, optimize battery performance, and extend the operational lifespan of LIBs for various applications.

Keywords: distribution of relaxation time; electrochemical impedance spectroscopy; degradation mechanism; solid electrolyte interface; loss of active material; loss of lithium ion

1. Introduction

The widespread adoption of electric vehicles (EVs) has driven extensive research into lithium-ion batteries (LIBs), which serve as the primary power source for these vehicles. Battery degradation is a focal point in this research due to its significant implications. As LIBs age, capacity degradation and increased internal resistance diminish both the driving range and power capacity of EVs, leading to suboptimal driving experiences [1,2]. Additionally, various side reactions occur within LIBs during aging, causing internal structural damage and raising the risk of thermal runaway, ultimately resulting in potential safety hazards [3,4]. Understanding the aging mechanisms and degradation models is crucial for estimating battery health based on historical data, optimizing current operating conditions, and predicting future performance.

Battery aging analysis encompasses various levels of investigation, including factors influencing degradation, internal side reactions, degradation modes, and external effects [5,6]. The most intuitive external characteristics of battery degradation are capacity fade and power fades [7,8]. At present, most papers still focus on these two points to conduct battery aging investigations and modeling. It should be noted that power fade is usually more difficult to investigate, so internal resistance is often analyzed instead [9]. Regarding degradation modes, battery aging mechanisms are often categorized as Loss of Lithium Ion (LLI) and loss of anode/cathode active materials for battery management and online diagnosis [10,11]. The charging and discharging processes of batteries are closely tied to lithium-ion intercalation and deintercalation on the active materials of the anode and cathode [12–14]. Consequently, battery capacity is directly influenced by the quantity of active materials and the availability of lithium ions, similar to how the capacity of a tank depends on its size and the amount of water it can hold. The primary degradation mechanisms of LIBs involve the loss of anode/cathode active materials and the LLI, analogous to water loss from the tank [15]. Additionally, the loss of electrolytes, including additives, is significant, with excessive loss potentially leading to capacity decline as batteries age.

Electrochemical Impedance Spectroscopy (EIS) is a valuable tool for studying material behavior over time, such as in corrosion, supercapacitors, and fuel cells [16–18]. EIS enables separate and sensitive analysis of each electrochemical reaction process without causing disruptions or damage [19]. EIS spectra offer valuable insights into LIB degradation [20]. However, analyzing EIS typically involves complex non-linear least squares fitting, which relies on the Equivalent Circuit Model (ECM). Finding suitable circuits for complex electrochemical phenomena poses challenges. Researchers are actively seeking alternative approaches to interpret EIS data. One promising method is the Distribution of Relaxation Time (DRT), which transforms data from the frequency (f) domain to the time (τ) domain. DRTs reveal peaks on either the time or frequency axis, indicating underlying electrochemical processes. The key advantage of DRTs is their model-free representation, requiring minimal system information. Various methods, such as Tikhonov Regularization (TR) [21,22], Maximum Entropy (ME) [23], and Fast Fourier Transform (FFT) [24,25], have been employed to convert impedance data to DRT. TR, for instance, involves adjusting regularization parameters for accurate DRTs, although different values can result in varying peak numbers, potentially obscuring true electrochemical phenomena. Despite this, TR does not require smoothing parameters or window functions, simplifying its application.

This research aims to enhance the analysis of the DRT for effectively separating overlapping time scales in EIS and providing clearer graphical interpretations [26,27]. Previous studies have primarily focused on capacity fade and impedance growth, but they often lack a sufficient analysis of DRT evolution in relation to cyclic aging and degradation mechanisms [28,29]. By utilizing an nth RC model with Gaussian functions, this study offers a detailed investigation into the behavior of characteristic DRT peaks and their correlation with battery aging. This approach fills gaps in existing methodologies, providing a more precise understanding of LIB degradation pathways and contributing to the development of advanced strategies for optimizing battery performance and management [30]. By directly correlating DRT peaks with specific battery degradation mechanisms, it offers novel insights into aging phenomena like SEI growth and electrode degradation. Practical validation with 18650 NMC lithium-ion batteries highlights its relevance for real-world battery diagnostics and performance monitoring.

The structure of this paper is as follows: Section 2 explains the application of the DRT method in interpreting EIS data, followed by a detailed explanation of deriving the DRT methods. The formulation of DRT equations based on Gaussian functions and regularization for the nth-order RC model is presented. Section 3 covers the acquisition of EIS data from experimental measurements, including frequency, real, and imaginary impedance components. Section 4 examines the correlation between the EIS Nyquist plot and the DRT plot, along with the interpretation of each peak in the DRT plots. It also explores the evolution of the EIS spectrum and DRT plots using data from 18650 batteries. The analysis highlights the identification of aging and degradation mechanisms by focusing on frequency-specific peaks in the DRT plots, particularly in the intermediate and low-frequency regions. The concluding section synthesizes the findings, summarizing insights and proposing future research directions.

2. Transformation of EIS Data into Distribution of Relaxation Time

2.1. Workflow of DRT Method

The analysis of the DRT method is structured into three main components, data preprocessing, discretization functions, and DRT calculation, as illustrated in Figure 1. Initially, EIS data $Z(\omega)$ is obtained through periodic checkup tests, which will be elaborated upon in Section 3. The data $Z(\omega)$ is classified based on the characteristics of its components. Following this categorization, the DRT of the electrochemical system is obtained through deconvolution. An integral aspect of this process involves discretizing the electrochemical impedance into a linear superposition function $G(\tau)$ using the weight function $X(m)$ and the radial base function.

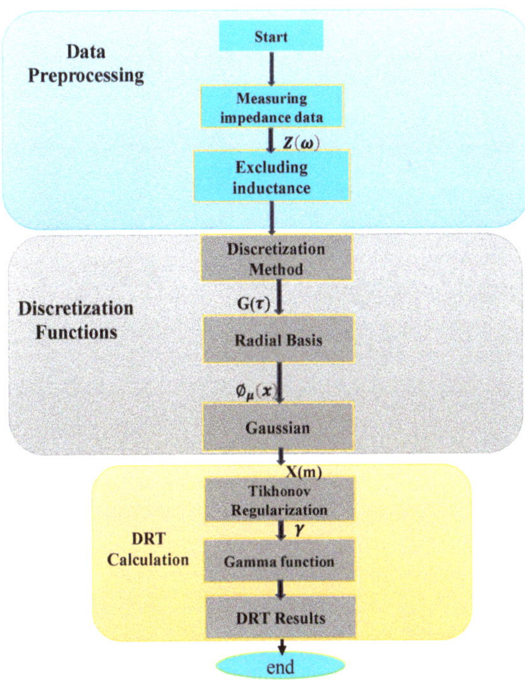

Figure 1. Flowchart of DRT calculation for measured EIS data.

It is important to note the inherent challenge posed by this process, requiring the regularization method to address the ill-posed nature of finding $\Phi(x)$. By employing ridge regression regularization, the weight function $X(m)$ is determined through quadratic programming to minimize the objective function, which represents the system matrix prior to optimization. Furthermore, the selection of the radial base function plays a critical role in minimizing approximation errors, as it integrates Gaussian functions across the frequency domain. The weight function $X(m)$ is then applied to the corresponding TR and linearly combined to produce the reconstructed DRT. A comprehensive breakdown of the detailed procedural steps and methodology is provided in the subsequent sections.

2.2. Derivation of DRT Method

In electrochemical research, the DRT method is a valuable tool for understanding relaxation processes in complex systems. This method excels at analyzing impedance across different frequencies, allowing researchers to determine the DRTs within electrochemical systems using specialized techniques [31]. The foundation of the DRT method lies in the concept that a complex frequency impedance spectrum can be represented as a composite of multiple distinct time constants, as illustrated in Figure 2, and the ECM of LIB with DRT peaks is shown in Figure 3.

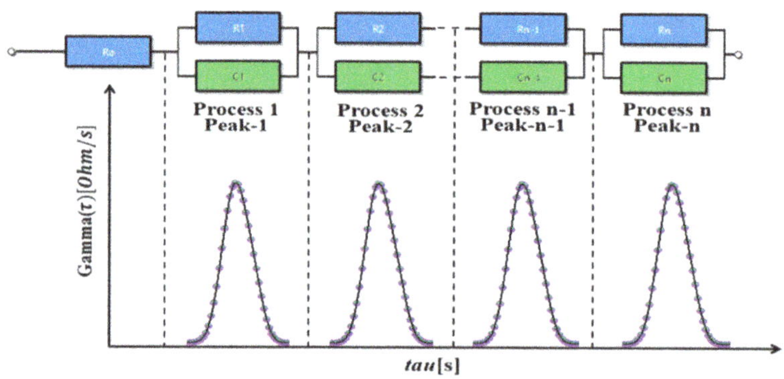

Figure 2. DRT peaks and general nth order R-C circuit model.

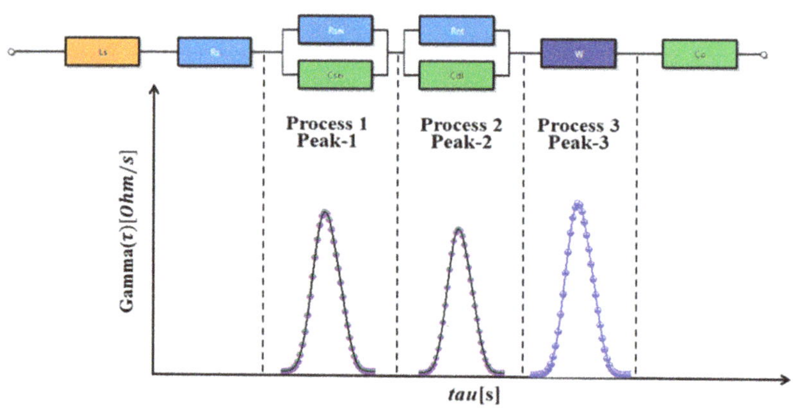

Figure 3. DRT peaks and an ECM of Li-ion batteries.

The impedance of an electrochemical system can therefore be expressed by (1)

$$Z(\omega) = R_0 + \int_0^\infty \frac{g(\tau)}{1 + j\omega\tau} d\tau \quad (1)$$

where R_0 is the ohmic resistance and it is the impedance at very high frequencies where the capacitive effects become negligible, $g(\tau)$ is a function that describes the DRTs within the system, ω is the angular frequency at which the impedance is measured, τ = RC represents the relaxation time where R and C note the effective resistance and capacitance, respectively.

The DRT is obtained by analyzing Figure 2, treating the impedance as a distribution of time constants.

$$Z(\omega) = R_0 + \int_{-\infty}^{+\infty} \frac{\gamma(ln\tau)}{1 + j\omega\tau} d(ln\tau) \tag{2}$$

where $\int_{-\infty}^{+\infty} \gamma(ln\tau)\mathrm{d}(ln\tau) = 1$

Equation (2) illustrates the relationship between impedance data and frequency distribution. Here, R_0 represents the ohmic resistance, $1/(1 + j\omega\tau)$ denotes the DRT kernel, and $\gamma(\ln\tau) = \tau g(\tau) \geq 0$ represents the distribution function of relaxation time.

Now using (2) and separating real and imaginary parts of impedance then equations will become

$$\mathrm{Zre}(\omega) = \mathrm{R}_0 + \int_{-\infty}^{+\infty} \frac{\gamma(\ln\tau)}{1 + \omega^2\tau^2} \mathrm{d}(\ln\tau) \tag{3}$$

$$\mathrm{Zimg}(\omega) = \int_{-\infty}^{+\infty} \frac{\gamma(\ln\tau)\omega\tau}{1 + \omega^2\tau^2} \mathrm{d}(\ln\tau) \tag{4}$$

DRT analysis aims to estimate the spectrum $\gamma(\ln\tau)$, which is an ill-posed problem, meaning it lacks unique solutions. To tackle this, discretization is applied to form a linear system of equations. Equations (3) and (4) are in continuous form but can be converted into a discrete form using $\tau_n = \frac{1}{2\pi f_n}, n = 1, 2, 3 \ldots N$, and N is the total number of frequency points. The resulting discrete forms are represented by (5) and (6), respectively.

$$Zre(\omega) = R_0 + \sum_{n=1}^{N} \frac{\gamma(ln(\tau))\Delta ln(\tau)}{1 + \omega^2\tau_n^2} \tag{5}$$

$$Zimg(\omega) = \sum_{n=1}^{N} \frac{\gamma(ln(\tau))\Delta ln(\tau)\omega\tau}{1 + \omega^2\tau_n^2} \tag{6}$$

Now expanding these equations up to n angular frequency $\omega_n, n = 1, 2, 3 \ldots n$, points and the impedance of series nth RC elements can be represented by the equation set as (7) and (8) for real and imaginary parts.

$$\left.\begin{array}{l} Zre(\omega_1) = R_0 + \frac{\gamma(ln(\tau_1))\Delta ln(\tau)}{1+\omega_1{}^2\tau_1^2} + \frac{\gamma(ln(\tau_2))\Delta ln(\tau)}{1+\omega_1{}^2\tau_2^2} + \ldots\ldots + \frac{\gamma(ln(\tau_n))\Delta ln(\tau)}{1+\omega_1{}^2\tau_n^2} \\ \vdots \\ Zre(\omega_n) = R_0 + \frac{\gamma(ln(\tau_1))\Delta ln(\tau)}{1+\omega_n{}^2\tau_1^2} + \frac{\gamma(ln(\tau_2))\Delta ln(\tau)}{1+\omega_n{}^2\tau_2^2} + \ldots\ldots + \frac{\gamma(ln(\tau_n))\Delta ln(\tau)}{1+\omega_n{}^2\tau_n^2} \end{array}\right\} \tag{7}$$

$$\left.\begin{array}{l} Zimg(\omega_1) = \frac{\gamma(ln(\tau_1))\Delta ln(\tau)\omega_1\tau_1}{1+\omega_1{}^2\tau_1^2} + \frac{\gamma(ln(\tau_2))\Delta ln(\tau)\omega_1\tau_2}{1+\omega_1{}^2\tau_2^2} + \ldots\ldots + \frac{\gamma(ln(\tau_n))\Delta ln(\tau)\omega_1\tau_n}{1+\omega_1{}^2\tau_n^2} \\ \vdots \\ Zimg(\omega_n) = \frac{\gamma(ln(\tau_1))\Delta ln(\tau)\omega_n\tau_1}{1+\omega_n{}^2\tau_1^2} + \frac{\gamma(ln(\tau_2))\Delta ln(\tau)\omega_n\tau_2}{1+\omega_n{}^2\tau_2^2} + \ldots\ldots + \frac{\gamma(ln(\tau_n))\Delta ln(\tau)\omega_n\tau_n}{1+\omega_n{}^2\tau_n^2} \end{array}\right\} \tag{8}$$

Two primary discretization methods for DRT analysis are the piecewise linear and Radial Basis Function (RBF) approaches. Among these, the RBF Gaussian method offers significant advantages due to its ability to approximate data with smooth and continuous Gaussian-shaped functions. This method ensures seamless transitions between adjacent relaxation processes, providing a more accurate resolution of overlapping peaks compared to the abrupt transitions inherent in the piecewise linear approach. Additionally, the RBF Gaussian method allows for greater control over the smoothness and resolution through parameters such as the Full Width at Half Maximum (FWHM) coefficient or shape factor, making it adaptable to complex datasets. The selection of Gaussian parameters, such as shape factor (μ) and FWHM, is crucial for resolving overlapping electrochemical processes in DRT analysis. A smaller μ improves resolution but increases noise sensitivity, while a larger μ smooths transitions but may oversimplify processes. Proper tuning ensures accurate separation of processes, while regularization techniques like Tikhonov mitigate noise and instability. By minimizing numerical artifacts and reducing noise amplification, this method delivers a more reliable interpretation of the relaxation time distribution. These attributes make the RBF Gaussian approach particularly suited to analyzing the intricate electrochemical behaviors of LIBs, as it captures subtle degradation mechanisms with high precision. In contrast, the piecewise linear method, while simpler, may oversimplify the data, potentially leading to inaccurate characterization of relaxation processes. Thus, the RBF Gaussian method emerges as the superior choice for DRT analysis, especially in studies focused on understanding battery aging and performance dynamics [32].

TR stabilizes DRT reconstruction by adding a term to the objective function, promoting smoother solutions. For RBF Gaussian discretization

$$\varnothing_{\mu}(x) = \exp\left(-(\mu x)^2\right) \tag{9}$$

where $x = |\ln\tau - \ln\tau_m|$, τ_m is the relaxation time of the mth collocation points and μ is the shape factor of the RBFs.

$$A_{re}(\omega) = \begin{bmatrix} \frac{1}{1+\omega_1{}^2\tau_1^2} & \frac{1}{1+\omega_1{}^2\tau_2^2} \cdots & \frac{1}{1+\omega_1{}^2\tau_n^2} \\ \vdots & \vdots & \vdots \\ \frac{1}{1+\omega_n{}^2\tau_1^2} & \frac{1}{1+\omega_n{}^2\tau_2^2} \cdots & \frac{1}{1+\omega_n{}^2\tau_n^2} \end{bmatrix} \tag{10}$$

$$A_{img}(\omega) = \begin{bmatrix} \frac{\omega_1\tau_1}{1+\omega_1{}^2\tau_1^2} & \frac{\omega_1\tau_2}{1+\omega_1{}^2\tau_2^2} \cdots & \frac{\omega_1\tau_n}{1+\omega_1{}^2\tau_n^2} \\ \vdots & \vdots & \vdots \\ \frac{\omega_n\tau_1}{1+\omega_n{}^2\tau_1^2} & \frac{\omega_n\tau_2}{1+\omega_n{}^2\tau_2^2} \cdots & \frac{\omega_n\tau_n}{1+\omega_n{}^2\tau_n^2} \end{bmatrix} \tag{11}$$

$$b_{re}(\omega) = \begin{bmatrix} Z_{re}(\omega_1) \\ \vdots \\ Z_{re}(\omega n) \end{bmatrix};\ b_{img}(\omega) = \begin{bmatrix} Z_{img}(\omega_1) \\ \vdots \\ Z_{img}(\omega n) \end{bmatrix} \tag{12}$$

The matrices A_{re} and A_{img} represent the real and imaginary components of impedance data while b_{re} and b_{img} represent the vectors containing the real and imaginary parts of the measured impedance data. When A_{re} and A_{img} matrices are combined, they yield matrix A, and the combination of b_{re} and b_{img} matrices result in matrix b. Consequently, in matrix notation, the overarching equation is expressed as Ax = b, where A belongs to $R^{m \times n}$, x belongs to R^n, and b belongs to R^m. Matrix A comprises m rows and n columns, while x is a matrix with n elements, and b is a matrix with m elements.

$$A = \begin{bmatrix} A_{re}(\omega) \\ A_{img}(\omega) \end{bmatrix} \quad b = \begin{bmatrix} b_{re}(\omega) \\ b_{img}(\omega) \end{bmatrix} \quad \text{and} \quad x = \begin{bmatrix} R_1 \\ R_2 \\ \vdots \\ R_n \\ R_0 \end{bmatrix} \tag{13}$$

From (13) and minimizing the solution equation will become as

$$min_x\left\{\|Ax - b\|^2 + \lambda\|x\|^2\right\} \tag{14}$$

The minimization of above equation has unique solution for any value of the regularization parameter $\lambda > 0$. The equation will be

$$x = \left(A^T A + \lambda^2 I\right)^{-1} A^T b \tag{15}$$

The higher λ value causes a stronger penalty to the sum of squares for fitting, which results in a smoother DRT profile. On the other hand, the smaller λ value causes the oscillation of the DRT profile.

The MEM is commonly used to transform EIS data into DRT due to its ability to produce smooth and stable solutions. However, MEM has significant disadvantages illustrated in Figure 4. One major drawback is its tendency to overly smooth the DRT plot, which can merge distinct electrochemical processes, such as charge transfer and diffusion, into a single feature. This smoothing effect makes it challenging to differentiate individual processes and analyze their relaxation times accurately.

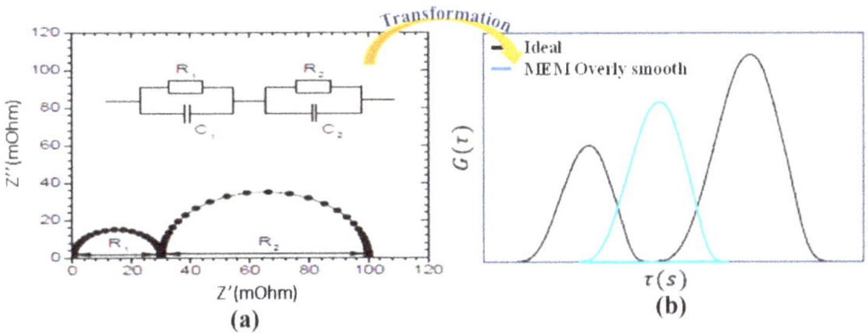

Figure 4. Transformation of EIS data using MEM. (**a**) Equivalent Semicircles (**b**) Transformed Peaks.

Additionally, MEM relies on prior assumptions that introduce bias into the results. This bias, combined with the loss of critical details in the DRT, limits MEM's suitability for battery diagnostics, where precise identification of fine structures is essential.

The FFT is also a widely used algorithm for transforming time domain data into the frequency domain due to its computational efficiency. However, FFT has significant drawbacks, particularly in noisy environments, as illustrated in Figure 5.

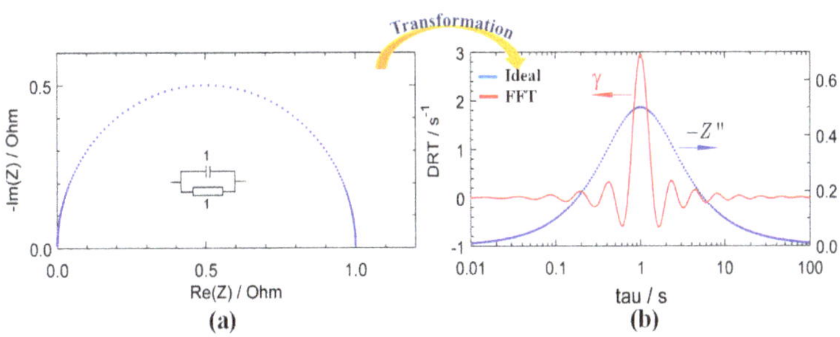

Figure 5. Transformation of EIS data using the FFT method. (**a**) Equivalent Semicircle (**b**) Transformed Peak.

FFT is highly sensitive to noise, where even small disturbances in the time-domain signal can distort frequency components, leading to multiple peaks in the results. For instance, Figure 5 shows multiple peaks in the FFT output, despite only one semicircle being present in the Nyquist plot. This makes it difficult to analyze true relaxation processes accurately.

Tikhonov Regularization (TR) is robust and stable as compared to methods such as ME and FFT. However, it also has practical challenges including the selection of optimal regularization parameters, sensitivity to noise, computational intensity, and reliance on smooth distribution assumptions. Despite its advantages, TR may face limitations in handling non-smooth cases and real-time applications.

3. Battery Aging Test Method

This section describes the aging test of a cylindrical 18650 LIB. The cylindrical battery cell used in the measurement has specifications provided in Table 1. The battery cell underwent aging through repeated charging and discharging cycles to generate EIS AC impedance spectra measured every five cycles. Changes in impedance parameters during the aging period were extracted using the measurement results and ECMs followed by curve fitting. The battery aging was conducted using HYSCLAB B.O.D Incubator equipment, maintaining a temperature of 25 °C. The charge and discharge tests as well as the EIS tests were conducted using WonATech WEIS-500. Cycle tests were performed with constant current charging at 0.5C and constant current discharging at 0.5C. Figure 6 illustrates the battery testing conditions and the evaluation of Nyquist plots, with a one-hour rest period after each charging and discharging cycle. EIS tests were conducted every five cycles, and impedance spectra were measured using a 60 mV perturbation in the frequency range of 0.1–1 kHz with the battery at a state of charge (SOC) of 100%. The cycle test was conducted up to 1035 cycles, during which the battery capacity decayed from 100% to 75%. This marked the point at which the battery could no longer function effectively, indicating it had reached the end of its usable life. The failure occurred due to an unknown reason, highlighting the need for further investigation into the underlying degradation mechanisms as illustrated in Figure 6 capacity graph. The selection of 18650 batteries was based on their widespread use in electric vehicles and consumer electronics, making them ideal for studying degradation mechanisms and ensuring the practical relevance of the analysis.

Table 1. Specification of the 18650 battery used for the aging test.

Property	Value
Chemistry	Nickel Manganese Cobalt
Type	18650
Capacity. max	2850 mAh
Nominal voltage	3.65 V

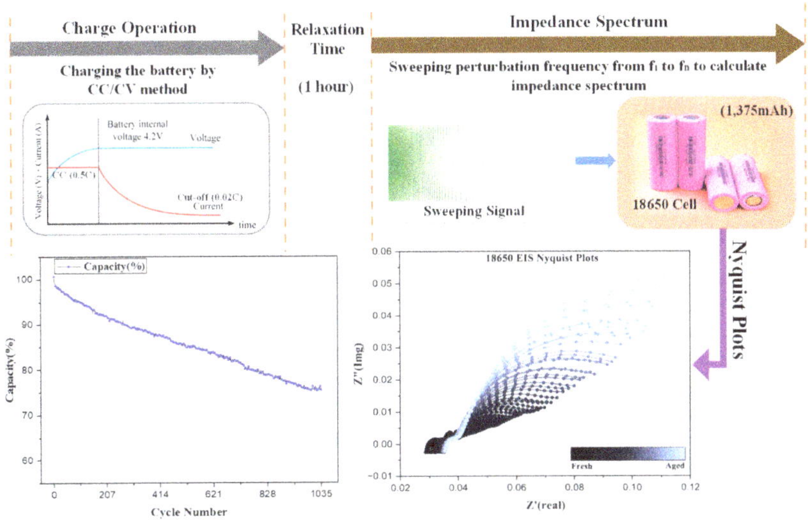

Figure 6. Battery testing conditions and evaluation of Nyquist plots.

4. Results and Analysis

4.1. Correlation Between EIS Nyquist Plots and DRT Plots

EIS Nyquist plot is a graphical representation where the real part of impedance (Z') is plotted on the *x*-axis, and the imaginary part (Z'') on the *y*-axis, offering insights into the electrical properties of a LIB, such as resistance, capacitance, charge transfer kinetics, and Warburg impedance. Each data point on the Nyquist plot corresponds to the impedance of the LIB at a specific frequency, with different regions of the plot providing information about various electrochemical processes as illustrated in Figure 7, such as the high-frequency region reflecting ohmic resistance and the lower frequency region indicating diffusion impedance.

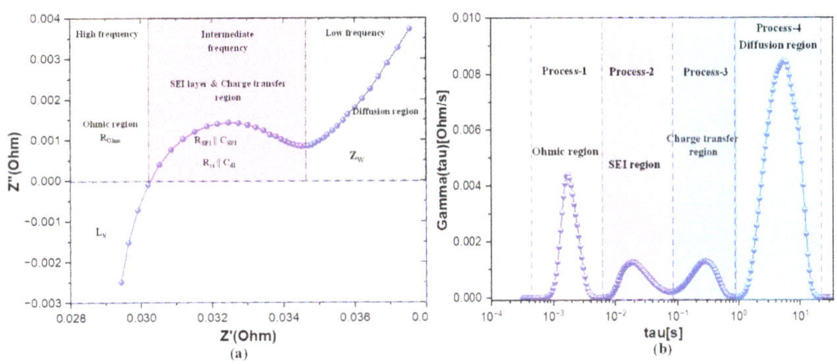

Figure 7. Transformation of an EIS plot to DRT plot for a 18650 cell (**a**) Nyquist plot of a 18650 cell (**b**) corresponding DRT plot transformed from the EIS plot.

In contrast, DRT analysis offers a model-free approach to interpreting EIS data by directly examining the DRTs within the system. DRT identifies and quantifies individual relaxation processes, with each peak in the DRT spectrum corresponding to a distinct electrochemical process, as shown in Figure 7b. This method provides clearer and more interpretable insights into the behavior of LIBs compared to the traditional approach of fitting EIS Nyquist plots to an ECM. The relationship between EIS Nyquist plots and

DRT plots is visually represented in Figure 7, where DRT peaks correspond to different features in the Nyquist plot, helping to connect and interpret the electrochemical processes more effectively.

EIS is a powerful tool for analyzing the internal processes of LIBs. However, in traditional Nyquist plots, overlapping semicircles often obscure individual electrochemical processes, making it challenging to differentiate and analyze the underlying mechanisms. To address this limitation, the DRT method is employed, transforming EIS data into a format where each electrochemical process is represented as a distinct peak. This transformation enables the clear separation and identification of individual processes such as the ohmic region, Solid Electrolyte Interphase (SEI) layer, charge transfer process, and diffusion process, providing deeper insights into the internal behavior of LIBs.

4.2. Detection of the Battery Status Using DRT Plot

Nyquist plots of the 18650 cell obtained from EIS measurements are presented in Figure 8a,b. Impedance spectrums were recorded for this cell after a specified number of cycles until the capacity decayed to 75%, marking the onset of failure. The spectrum is divided into four frequency ranges: ultra-high frequency, high frequency, intermediate frequency, and low frequency. The semicircle at the intermediate frequency is associated with charge transfer, while the low-frequency region corresponds to diffusion processes, both closely related to battery degradation during cycling. In contrast, the ultra-high and high-frequency regions are associated with ohmic resistance and the SEI region, respectively, which are less affected by cycling. In Figure 8a,b, overlapping semicircles make it challenging to distinguish individual electrochemical processes. Ideally, each process would correspond to a distinct semicircle, but in practice, they often overlap. Additionally, the height and width of the semicircles increase with cycling, reflecting changes in the impedance spectrum. However, due to similarities in the electrochemical processes, these semicircles overlap, complicating the derivation of individual relaxation times and amplitudes.

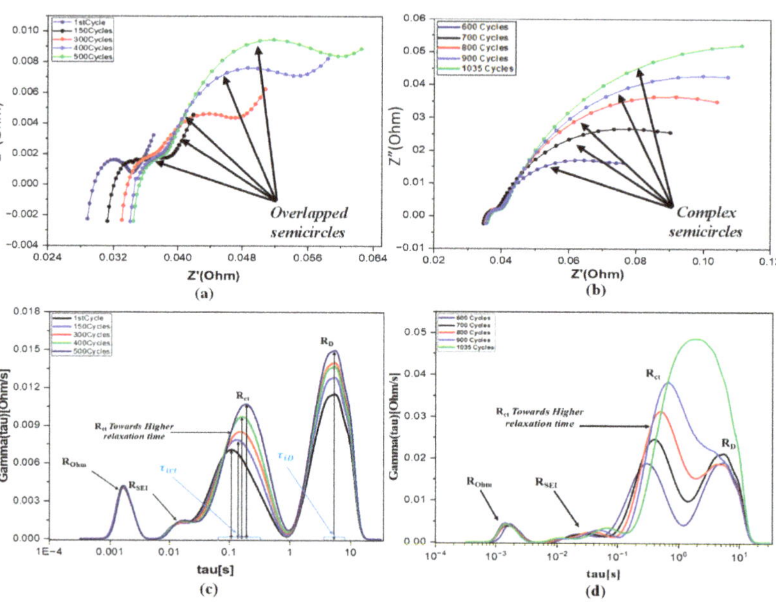

Figure 8. (**a**,**b**) Nyquist plots of the 18650 cell over the life cycle (**c**,**d**) DRT plots of the 18650 cell over the life cycle.

There is an ongoing need in battery research for methods that combine frequency-domain and time-domain analysis to interpret EIS for understanding battery characteristics and diagnosing issues [33]. DRT analysis addresses this need by enabling a clearer breakdown and understanding of each process compared to frequency-domain analysis alone. A broader range of frequencies provides detailed insights into internal processes within the battery. The higher-frequency region reveals faster processes such as chemical reactions, while the lower-frequency region highlights slower processes like diffusion of ions. Time-domain analysis also provides a clearer view of each electrochemical process, enhancing the understanding of battery behavior.

In Figure 8c,d, the DRT plot for the 18650 battery is shown up to 1035 cycles. Four peaks are observed in the DRT plot. The rightmost peak, which increases continuously with cycling, is referred to as the diffusion process. It indicates battery aging and is located at a higher relaxation time, corresponding to the lower frequency region [34]. The peak on the left side is known as the contact impedance or ohmic impedance, resulting from the direct physical contact between the active materials and the collectors. This peak shows almost no variation as the battery cycles, which is attributed to the high-quality production and manufacturing process of the anode material in LIBs. The middle two peaks are associated with the SEI layer and charge transfer process. As shown in Figure 8c,d, during the initial cycles, the SEI layer grows rapidly and then stabilizes. The charge transfer impedance increases with cycling and its shift toward higher relaxation times indicates that the process is slowing down. When the peaks of the charge transfer and diffusion processes merge at a single time constant tau, it signifies a critical state of degradation in the battery. The merging of these processes at a single-time constant tau indicates that the battery can no longer sustain electrochemical reactions. This is due to SEI layer formation, lithium plating, and loss of active materials, which increase internal resistance. This increase in impedance with cycling reflects the battery degradation mechanism.

The degradation mechanisms of LIBs can be thoroughly analyzed and better understood with the help of Figure 9 [35], which provides a comprehensive overview of the various processes involved. An increase in ohmic resistance also known as contact resistance in LIBs is linked to reduced electrical conductivity within the battery components, which can arise from mechanisms like electrolyte consumption and electrolyte decomposition. Electrolyte consumption leads to a decrease in the availability of charge carriers (ions), while electrolyte decomposition generates chemical species that might have lower conductivity. These factors collectively contribute to an increase in ohmic resistance, resulting in higher internal resistance and diminished battery performance. Additionally, an increase in the resistance of the SEI layer indicates a degradation process in LIBs [36]. The SEI layer, formed on the electrode surfaces due to electrolyte decomposition, suggests the LLIs within the battery, leading to a reduction in its capacity and performance over time [35].

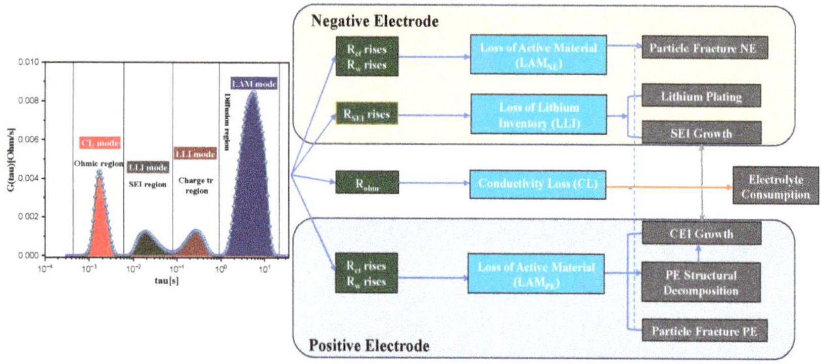

Figure 9. LIBs degradation mechanisms based on DRT plots.

Moreover, a rise in charge transfer and diffusion processes within LIBs may suggest a decrease in active material within the electrodes, particularly noticeable in the negative electrode (anode) and positive electrode (cathode). This impedance increase often signifies degradation processes that lead to reduced electron and ion transport within the electrodes, which can be attributed to particle fractures and structural changes in the active material as shown in Figure 9. One of the primary causes of such degradation is the phenomenon of graphite exfoliation and island formation in the negative electrode, typically made of graphite [37]. During charging and discharging cycles, the anode undergoes expansion and contraction due to the insertion and extraction of lithium ions, contributing to degradation over time [38]. This degradation can include particle cracking, LAM from the electrode, and changes in the crystal structure, ultimately reducing the capacity to store and release lithium ions.

5. Conclusions

This study demonstrates the deconvolution of EIS data into the DRT using TR with Gaussian functions, providing a more accurate and precise analysis of LIBs degradation mechanisms. By utilizing Gaussian functions, the DRT method effectively separates overlapping electrochemical processes and offers clearer insights into battery aging and performance decline compared to traditional methods. The analysis shows that shifts in electrochemical processes, such as charge transfer and diffusion, correlate with the degradation of LIBs over time. This approach offers a detailed understanding of the degradation mechanisms, including the role of the SEI layer in affecting battery capacity. The DRT method enhances the ability to track and quantify these changes more precisely, offering a clearer picture of battery health. The DRT method offers distinct advantages over traditional EIS analysis by deconvoluting overlapping electrochemical processes in Nyquist plots, providing higher resolution and greater insight into individual mechanisms such as SEI growth, charge transfer, and diffusion. Unlike conventional EIS, which often struggles with interpreting complex impedance spectra, the DRT method isolates and quantifies specific processes, thereby enabling more accurate analysis of battery aging and degradation. TR with Gaussian functions for DRT deconvolution proves to be a powerful tool for improving the accuracy of battery performance predictions. This method supports the development of advanced battery management systems by enabling better monitoring of battery health, optimizing performance, and extending the operational lifespan of LIBs across various applications. By providing a deeper understanding of battery degradation, it contributes to the advancement of more efficient and reliable battery technologies.

Author Contributions: Conceptualization, W.C. and M.S.; methodology, M.S.; software, M.S.; validation, W.C., M.S. and A.S.A.; formal analysis, M.S.; investigation, M.S.; resources, W.C.; data curation, M.S.; writing—original draft preparation, M.S.; writing—review and editing, A.S.A.; visualization, M.S. and A.S.A.; supervision, W.C.; project administration, W.C.; funding acquisition, W.C. All authors have read and agreed to the published version of the manuscript.

Funding: This research received no external funding.

Data Availability Statement: Data is unavailable due to privacy or ethical restrictions.

Conflicts of Interest: The authors declare no conflict of interest.

References

1. Abellan, P.; Mehdi, B.L.; Parent, L.R.; Gu, M.; Park, C.; Xu, W.; Zhang, Y.; Arslan, I.; Zhang, J.-G.; Wang, C.-M.; et al. Probing the Degradation Mechanisms in Electrolyte Solutions for Li-ion Batteries by In-Situ Transmission Electron Microscopy. *Nano Lett.* **2014**, *14*, 1293–1299. [CrossRef] [PubMed]
2. Shchurov, N.I.; Dedov, S.I.; Malozyomov, B.V.; Shtang, A.A.; Martyushev, N.V.; Klyuev, R.V.; Andriashin, S.N. Degradation of lithium-ion batteries in an electric transport complex. *Energies* **2021**, *14*, 8072. [CrossRef]
3. Golubkov, A.W.; Fuchs, D.; Wagner, J.; Wiltsche, H.; Stangl, C.; Fauler, G.; Voitic, G.; Thaler, A.; Hacker, V. Thermal-runaway experiments on consumer Li-ion batteries with metal-oxide and olivin-type cathodes. *Rsc Adv.* **2014**, *4*, 3633–3642. [CrossRef]
4. Zaghib, K.; Dontigny, M.; Guerfi, A.; Trottier, J.; Hamel-Paquet, J.; Gariepy, V.; Galoutov, K.; Hovington, P.; Mauger, A.; Groult, H.; et al. An improved high-power battery with increased thermal operating range: C–$LiFePO_4$//C–$Li_4Ti_5O_{12}$. *J. Power Sources* **2012**, *216*, 192–200. [CrossRef]
5. Birkl, C.R.; Roberts, M.R.; McTurk, E.; Bruce, P.G.; Howey, D.A. Degradation diagnostics for lithium ion cells. *J. Power Sources* **2017**, *341*, 373–386. [CrossRef]
6. Sohaib, M.; Choi, W. *Investigation of the Aging Phenomena in Lithium-Ion Batteries Using Distribution of Relaxation Time Analysis*; The Korean Institute of Power Electronics: Seoul, Republic of Korea, 2024.
7. Bloom, I.; Cole, B.; Sohn, J.; Jones, S.; Polzin, E.; Battaglia, V.; Henriksen, G.; Motloch, C.; Richardson, R.; Unkelhaeuser, T.; et al. An accelerated calendar and cycle life study of Li-ion cells. *J. Power Sources* **2001**, *101*, 238–247. [CrossRef]
8. Wright, R.B.; Christophersen, J.P.; Motloch, C.G.; Belt, J.R.; Ho, C.D.; Battaglia, V.S.; Barnes, J.A.; Duong, T.Q.; Sutula, R.A. Power fade and capacity fade resulting from cycle-life testing of Advanced Technology Development Program lithium-ion batteries. *J. Power Sources* **2003**, *119–121*, 865–869. [CrossRef]
9. Belt, J.; Utgikar, V.; Bloom, I. Calendar and PHEV cycle life aging of high-energy, lithium-ion cells containing blended spinel and layered-oxide cathodes. *J. Power Sources* **2011**, *196*, 10213–10221. [CrossRef]
10. Pop, V.; Bergveld, H.J.; Regtien, P.P.L.; het Veld, J.O.; Danilov, D.; Notten, P.H.L. Battery aging and its influence on the electromotive force. *J. Electrochem. Soc.* **2007**, *154*, A744e50. [CrossRef]
11. Sarasketa-Zabala, E.; Aguesse, F.; Villarreal, I.; Rodriguez-Martinez, L.M.; López, C.M.; Kubiak, P. understanding lithium inventory loss and sudden performance fade in cylindrical cells during cycling with deep-discharge steps. *J. Phys. Chem. C* **2015**, *119*, 896–906. [CrossRef]
12. Wang, X.; Wei, X.; Chen, Q.; Zhu, J.; Dai, H. Lithium-ion battery temperature on-line estimation based on fast impedance calculation. *J. Energy Storage* **2019**, *26*, 100952. [CrossRef]
13. Al Nazer, R.; Cattin, V.; Granjon, P.; Montaru, M.; Ranieri, M. broadband identification of battery electrical impedance for HEVs. *IEEE Trans. Veh. Technol.* **2013**, *62*, 2896–2905. [CrossRef]
14. Gantenbein, S.; Weiss, M.; Ivers-Tiffée, E. Impedance based time-domain modeling of lithium-ion batteries: Part I. *J. Power Sources* **2018**, *379*, 317–327. [CrossRef]
15. Dubarry, M.; Truchot, C.; Liaw, B.Y. Synthesize battery degradation modes via a diagnostic and prognostic model. *J. Power Sources* **2012**, *219*, 204–216. [CrossRef]
16. Encinas-Sánchez, V.; de Miguel, M.T.; Lasanta, M.I.; García-Martín, G.; Pérez, F.J. Electrochemical impedance spectroscopy (EIS): An efficient technique for monitoring corrosion processes in molten salt environments in CSP applications. *Sol. Energy Mater. Sol. Cells* **2019**, *191*, 157–163. [CrossRef]
17. Sakita, A.M.P.; Della Noce, R.; Lavall, R.L. Potential-dependent electrochemical impedance spectroscopy as a powerful tool for evaluating supercapacitor electrode performance. *J. Electrochem. Soc.* **2021**, *168*, 080525. [CrossRef]
18. Han, S.B.; Oh, H.; Lee, W.Y.; Won, J.; Chae, S.; Baek, J. On-line eis measurement for highpower fuel cell systems using simulink real-time. *Energies* **2021**, *14*, 6133. [CrossRef]
19. Magar, H.S.; Hassan, R.Y.A.; Mulchandani, A. Electrochemical impedance spectroscopy (Eis): Principles, construction, and biosensing applications. *Sensors* **2021**, *21*, 6578. [CrossRef] [PubMed]

20. Nara, H.; Yokoshima, T.; Osaka, T. Technology of electrochemical impedance spectroscopy for an energy-sustainable society. *Curr. Opin. Electrochem.* **2020**, *20*, 66–77. [CrossRef]
21. Ivers-Tiffée, E.; Weber, A. Evaluation of electrochemical impedance spectra by the distribution of relaxation times. *J. Ceram. Soc. Jpn.* **2017**, *125*, 193–201. [CrossRef]
22. Weiß, A.; Schindler, S.; Galbiati, S.; Danzer, M.A.; Zeis, R. Distribution of relaxation times analysis of high-temperature PEM fuel cell impedance spectra. *Electrochim. Acta* **2017**, *230*, 391–398. [CrossRef]
23. Hörlin, T. Deconvolution and maximum entropy in impedance spectroscopy of noninductive systems. *Solid State Ion.* **1998**, *107*, 241–253. [CrossRef]
24. Boukamp, B.A. Derivation of a Distribution Function of Relaxation Times for the (fractal) Finite Length Warburg. *Electrochim. Acta* **2017**, *252*, 154–163. [CrossRef]
25. Boukamp, B.A.; Rolle, A. Analysis and application of distribution of relaxation times in solid state ionics. *Solid State Ion.* **2017**, *302*, 12–18. [CrossRef]
26. Qahouq, J.A.A.; Xia, Z. Single-perturbation-cycle online battery impedance spectrum measurement method with closed-loop control of power converter. *IEEE Trans. Ind. Electron.* **2017**, *64*, 7019–7029. [CrossRef]
27. Dam, S.K.; John, V. High-resolution converter for battery impedance spectroscopy. *IEEE Trans. Ind. Appl.* **2018**, *54*, 1502–1512. [CrossRef]
28. Gaberšček, M. Understanding Li-based battery materials via electrochemical impedance spectroscopy. *Nat. Commun.* **2021**, *12*, 19–22. [CrossRef] [PubMed]
29. Shafiei Sabet, P.; Warnecke, A.J.; Meier, F.; Witzenhausen, H.; Martinez-Laserna, E.; Sauer, D.U. Non-invasive yet separate investigation of anode/cathode degradation of lithium-ion batteries (nickel–cobalt–manganese vs. graphite) due to accelerated aging. *J. Power Sources* **2020**, *449*, 227369. [CrossRef]
30. Xiong, R.; Li, L.; Tian, J. Towards a smarter battery management system: A critical review on battery state of health monitoring methods. *J. Power Sources* **2018**, *405*, 18–29. [CrossRef]
31. Li, X.; Ahmadi, M.; Collins, L.; Kalinin, S.V. Deconvolving distribution of relaxation times, resistances and inductance from electrochemical impedance spectroscopy via statistical model selection: Exploiting structural-sparsity regularization and data-driven parameter tuning. *Electrochim. Acta* **2019**, *313*, 570–583. [CrossRef]
32. Huang, J.; Sullivan, N.P.; Zakutayev, A.; O'hayre, R. How reliable is distribution of relaxation times (DRT) analysis? A dual regression-classification perspective on DRT estimation, interpretation, and accuracy. *Electrochim. Acta* **2023**, *443*, 141879. [CrossRef]
33. Zhu, J.; Knapp, M.; Liu, X.; Yan, P.; Dai, H.; Wei, X.; Ehrenberg, H. Low-temperature separating lithium-ion battery interfacial polarization based on distribution of relaxation times (DRT) of impedance. *IEEE Trans. Transp. Electrif.* **2020**, *7*, 410–421. [CrossRef]
34. Sohaib, M.; Choi, W. *A Novel Aging Diagnosis Method Using Distribution of Relaxation Time*; The Korean Institute of Power Electronics: Seoul, Republic of Korea, 2024.
35. Huang, S.; Li, K.; He, Z.; Wang, Y.; Li, C.; Li, H.; Yu, H.; Chen, Y.; Chen, L. Rolling strategy for highly efficient preparation of phosphating interface enabled the stable lithium anode. *J. Alloys Compd.* **2024**, *1005*, 176193. [CrossRef]
36. Zhang, Y.; Tang, Q.; Zhang, Y.; Wang, J.; Stimming, U.; Lee, A.A. Identifying degradation patterns of lithium ion batteries from impedance spectroscopy using machine learning. *Nat. Commun.* **2020**, *11*, 1706. [CrossRef] [PubMed]
37. Pender, J.P.; Jha, G.; Youn, D.H.; Ziegler, J.M.; Andoni, I.; Choi, E.J.; Heller, A.; Dunn, B.S.; Weiss, P.S.; Penner, R.M.; et al. Electrode degradation in lithium-ion batteries. *ACS Nano* **2020**, *14*, 1243–1295. [CrossRef]
38. Yang, H.; Zheng, H.; Yu, H.; Qu, B.; Chen, L.; Niu, J.; Chen, Y. Coordinating ionic and electronic conductivity on 3D porous host enabling deep dense lithium deposition toward high-capacity lithium metal anodes. *Nanoscale* **2022**, *14*, 13722–13730. [CrossRef] [PubMed]

batteries

MDPI

Article

One-Dimensional Electro-Thermal Modelling of Battery Pack Cooling System for Heavy-Duty Truck Application

Mateusz Maciocha [1], Thomas Short [1], Udayraj Thorat [1], Farhad Salek [2], Harvey Thompson [1] and Meisam Babaie [1,*]

[1] School of Mechanical Engineering, University of Leeds, Leeds LS2 9JT, UK; u.y.thorat@leeds.ac.uk (U.T.); h.m.thompson@leeds.ac.uk (H.T.)
[2] AVL Powertrain UK LTD, Coventry CV4 7EZ, UK; farhad.salek@avl.com
* Correspondence: m.babaie@leeds.ac.uk

Abstract: The transport sector is responsible for nearly a quarter of global CO_2 emissions annually, underscoring the urgent need for cleaner, more sustainable alternatives such as electric vehicles (EVs). However, the electrification of heavy goods vehicles (HGVs) has been slow due to the substantial power and battery capacity required to match the large payloads and extended operational ranges. This study addresses the research gap in battery pack design for commercial HGVs by investigating the electrical and thermal behaviour of a novel battery pack configuration using an electro-thermal model based on the equivalent circuit model (ECM). Through computationally efficient 1D modelling, this study evaluates critical factors such as cycle ageing, state of charge (SoC), and their impact on the battery's range, initially estimated at 285 km. The findings of this study suggest that optimal cooling system parameters, including a flow rate of 18 LPM (litres per minute) and actively controlling the inlet temperature within ±7.8 °C, significantly enhance thermal performance and stability. This comprehensive electro-thermal assessment and the advanced cooling strategy set this work apart from previous studies centred on smaller EV applications. The findings provide a foundation for future research into battery thermal management system (BTMS) design and optimised charging strategies, both of which are essential for accelerating the industrial deployment of electrified HGVs.

Keywords: driving cycle; electric truck; dynamic battery thermal model; BTMS; equivalent circuit model

1. Introduction

One of the main challenges that the modern world continues to face is climate change and the global warming process associated with it. Global temperatures have increased by more than 1 °C since 1880, with most of this rise occurring since 1975 [1]. The main driver of this continues to be the global production of CO_2, with the transport sector accounting for 24% of the 33.5 billion tonnes produced annually [2]. Having been highlighted as a leading contributor, governments globally have placed stricter emissions legislation on the transport sector. The EU and UK governments have introduced policies to suspend the sale of fossil fuel-powered consumer vehicles beyond 2035 [3,4]. This will leave a void in sales of more than 10 million vehicles annually [5], which will likely be filled by vehicles featuring either electric or hydrogen propulsion systems. In anticipation of the transition to more environmentally friendly vehicles, the popularity of EVs (electric vehicles) has risen exponentially over recent years, with this trend likely to continue [6].

Despite similar legislation [7], the heavy-duty commercial vehicle sector has been slower to electrification than other sectors [8]. As of 2021, only 0.3% of global new truck registrations were attributed to electrically powered vehicles, accounting for less than 0.1% of the total global fleet [9]. The slow transition is likely due to significantly increased payloads [10] requiring advancements in batteries and electric drive units (EDUs).

EDUs capable of supplying the required power for electric heavy-duty trucks have now been developed, which can provide high power such as 375 kW continuously [11]. Despite this, limited research and development have been carried out to enable the production of equally capable battery systems. Where the literature is available, it usually focuses on passenger and light-duty commercial vehicles, hindering heavy-good vehicle electrification [11].

Automobile applications feature significantly more demanding technical requirements than customer electronics regarding capacity, operating temperature range, cost, and cycle life. Mathematical models can predict system characteristics in different environments and conditions, enabling assessment of the performance and range of EVs [12]. Due to the complexity of large-scale energy storage systems and their loading cycles, computational methods are most used for these assessments. Tools such as AVL Cruise and MATLAB Simulink R2023b allow for incorporating vehicle parameters, drive cycle requirements, and battery characteristics through mathematical equations representing physical and chemical reactions [13–15].

Mathematical models estimate the performance of individual cells or whole systems, with SoC, voltage, current draw, and power supply among the parameters of significant interest [15–17]. It is, therefore, essential to model these accurately to ensure physical battery operation is as predicted, as this allows manufacturers to establish whether performance targets can be met reliably and safely by the system. The ability to mathematically model lithium-ion (Li-ion) batteries is vital for the development of battery management systems (BMS), which help to maintain electrical and thermal balance in energy storage systems (ESS) [18].

Effective Battery Thermal Management Systems (BTMS) are needed within EV battery packs as Li-ion batteries' performance and safety characteristics are sensitive to cell temperatures [18]. It is generally accepted that Li-ion batteries tend to perform optimally between 25 °C and 35 °C [19,20]. Above this range, the cycle life of Li-ion batteries tends to decrease exponentially. Tarascone et al. [21] found that in early rechargeable Li-ion batteries, the battery cycle life at 25 °C was approximately 250% greater than that at 55 °C. Similar harmful impacts on Li-ion cell capacity have been found below optimal temperatures, with Lo [22] showing up to 30% reductions in cell capacity at −20 °C. Temperature differences within Li-ion cells and across battery packs have also been found to hinder overall performance. It has been suggested that a maximum difference of 5 °C should not be exceeded within cells and between modules to maintain optimal performance [23,24] and avoid potential short-circuiting [25]. Temperature differences more significant than this have increased thermal ageing by 25% and decreased power capabilities by 10% [26]. Based on existing studies, it can therefore be suggested that the BTMS must maintain optimal battery temperature (25–35 °C), with a pack temperature range of <5 °C.

Key thermal challenges facing the electrification of HGVs stem from the increased mass and power requirements compared to typical EVs. Typical EVs have masses of around 1600–2300 kg [27,28] and electric drive units capable of supplying 100–150 kW [28]. Due to payloads of up to 44 tonnes, HGVs require motors that can provide high power and torque to perform similarly to current diesel powertrains [29]. The additional mass of loaded HGVs means they can require upwards of 1000% more torque during standardised drive cycles than typical EVs [30–32], thus drawing proportionately more current from the battery pack. With total heat generation within Li-ion battery systems proportional to the amperage draw squared [24], the increased torque requirements mean that the heat generation per battery cell could be exponentially more significant in electric HGVs than typical EVs, increasing demand on the BTMS. Electric HGVs' more significant power requirements also mean they will likely have much larger battery systems than typical EVs. Typical EVs feature battery capacities of around 40–80 kWh [28], compared to an electric HGV requirement of upwards of 400 kWh [33], to achieve ranges competitive with existing diesel counterparts. At the same current draw, this may mean that the electric truck's total battery heat generation could be up to 750% greater than typical EVs. To maintain thermal stability within the battery, the truck's BTMS and cooling capacity would need to be upscaled sufficiently, with components such as the radiator, compressor, and pumps capable of generating increased heat.

According to Pesaran [34], EV BTMS should use minimal parasitic power while allowing pack operation in hot and cold climates. Potential BTMS designs for typical EVs have been compared in studies, with liquid water/glycol cooling widely regarded as preferential [35]. This has been attributed to the more significant heat transfer coefficients of liquid systems, which can be more than 3× those of equivalent air systems [34]. Such liquid BTMSs can be cooled passively and actively, as detailed by Piao et al. [36], with passive systems only found to be viable for climates between 10 °C and 35 °C [34].

The existing literature has widely studied potential liquid BTMS for consumer EVs, with the inlet temperature and coolant flow rate having been found to play critical roles. Lan et al. [37] studied the effect of coolant flow rate and detailed battery cell thermal behaviour at different C-rates. It was found that an increase in coolant flow rate provided increased cooling across all C-rates but would eventually suffer from diminishing returns. Karimi et al. [38] carried out similar work and achieved flow rate results identical to those of Lan et al. [37], but also studied the effect of inlet temperature on cell temperature. It was shown that a decrease in inlet temperature proportionately decreased maximal cell temperature across the whole battery discharging cycle. For example, at the end of the 1400 s cycle, each 1 °C reduction in coolant inlet temperature reduced the maximal end temperature by 0.6 °C. This trend was repeated across all inlet temperatures. Yue et al. [39] also found similar results at the pack level. These studies suggest that increased coolant flow rate and low coolant inlet temperatures would provide preferential cooling for electric HGV TMS. However, the diminishing returns shown by increased flow rates and the increased energy consumption required for active cooling suggest that both properties should be optimised for each application.

This study addresses these challenges by employing a 1D electro-thermal model within the MATLAB Simulink R2023b environment to assess a potential battery design's electrical and thermal performance for heavy-duty vehicles. This model is based on realistic cell characteristics, real-world power demand data, and driving cycle data for HGVs. Using a 1D model offers a practical solution by increasing the speed and reducing the costs of large-scale simulations while maintaining accuracy. Unlike previous research, this study integrates real-world power demand and drive cycle data for heavy-duty trucks. It also employs realistic cell characteristics and cooling system performance, ensuring the results apply to industrial-scale deployments. An equivalent circuit model (ECM) is calibrated for electro-thermal analysis using actual battery test data. This enhances the reliability of the simulation results, allowing for precise evaluation of battery behaviour under various operating conditions [40].

The primary novel contributions of the present research are as follows:

1. Thermal management solutions for heavy-duty electric vehicles, providing analysis of these vehicles' thermal behaviour.
2. Use of actual drive cycle and power demand data for realistic battery design.
3. The 1D electro-thermal modelling is based on ECM, battery test data, and typical thermal management materials.
4. Industrial-scale applicability using realistic cell characteristics and cooling system performance.

The structure of this paper is as follows: Section 2 outlines the methodology, covering the system description, electrical battery model, and thermal cooling model. Section 3 presents the study's findings, while Section 4 provides an in-depth discussion. Finally, Section 5 concludes with a summary of the key insights.

2. Materials and Methods

2.1. System Description

Prior to the modelling procedure, a battery pack design for an electric heavy-duty truck was created based on typical EV battery specifications and space availability in heavy-duty trucks. Samsung SDI 94 Ah cells [41] were selected for use in the battery pack design due to their suitability in automotive applications, historically proven by their presence in multiple EVs. These cells use lithium-ion chemistry, with a Nickel–Cobalt–Manganese (NMC)-based cathode composition ($Ni_{0.33}Mn_{0.33}Co_{0.33}O_2$) [42].

Following a review of the limited EV trucks available from large original equipment manufacturers (OEMs) [43], minimum electrical capabilities of the novel battery pack were established, with an excess of 600 V and 516 kWh required. Table 1 highlights the battery's electrical architecture, with Table 2 highlighting the system's electrical characteristics.

Table 1. Battery architecture used in the current study.

Type	Module	Pack	Total
Electrical architecture	12 cells in series (12 s)	16 modules in series	7 packs in parallel

Table 2. Battery system parameters used in the current study.

	Cell [41]	Module	Pack	Total
Nominal voltage (V)	3.68	44.16	706.56	706.56
Maximum voltage (V)	4.15	49.80	796.80	796.80
Capacity (Ah)	94	94	94	658
Energy capacity (kWh)	0.35	66.41	64.41	464.92
Maximum energy capacity (kWh)	0.39	74.52	74.52	521.64
Mass (kg)	2.1	28	448	3136

At the point of concept, consideration was made to the spatial requirements of such a battery to allow for its potential future integration into an HGV. An example of the novel battery in situ is shown by 3D visualisation in Figure 1. Prior work has shown that a cooling channel design is the most appropriate for a vehicle of this scale when considering cooling performance, pressure loss, and special requirements [44]. The cooling system was assumed to feature 6 cooling channels per pack, as shown in Figure 2, while using an industry-standard 50:50 water/glycol coolant mixture [45].

Figure 1. Rendering of the battery pack in the truck with and without aero covers.

Figure 2. 3D CAD model of battery module with cooling channels.

2.2. Battery Thermal Model

2.2.1. Battery Life Cycle Dynamics and Electrical Parameters Overview

Physical testing data of typical EV battery modules containing Samsung SDI prismatic cells was used to understand cell cycle ageing, with the results summarised in Table 3. The study considered 2000 and 4000 cycles, corresponding to 100%, 90%, and 80% of the state of health (SoH). Further cycle ageing was not considered due to the cell's end of life (EOL) being exceeded, with a replacement expected when the SoH of the cells depletes beyond 80%.

Table 3. Battery life cycle values of a typical EV battery module.

Charge/Discharge Cycles	Capacity Difference at 25 °C (%)	Internal Resistance Difference at 25 °C (%)
1	0	0
2000	−9.96	17.28
4000	−20.28	42.45

Key battery parameters were parametrised within the script and applied to each cell in line with manufacturer datasheets and experimental fade data. The complete block diagram of the electrical system, alongside charge/discharge application and battery performance assessment methods, is also presented in Figure 3. The power demand for each cycle was proportioned to each pack. This meant that including all 7 packs was optional, helping simplify the model while significantly decreasing the computational power without compromising accuracy. Uniform discharge and health degradation were assumed for all the cells in the system; hence, a SoC reading from the output of a single cell was considered sufficient. In the initial stage, the thermal implications were neglected, with perfect cooling assumed, therefore maintaining the cells' temperature at 298.15 K. The electrical model would later be combined with the thermal model to investigate the thermal impacts of varying power demand. Similarly, no self-discharge or memory effect was considered for the cells.

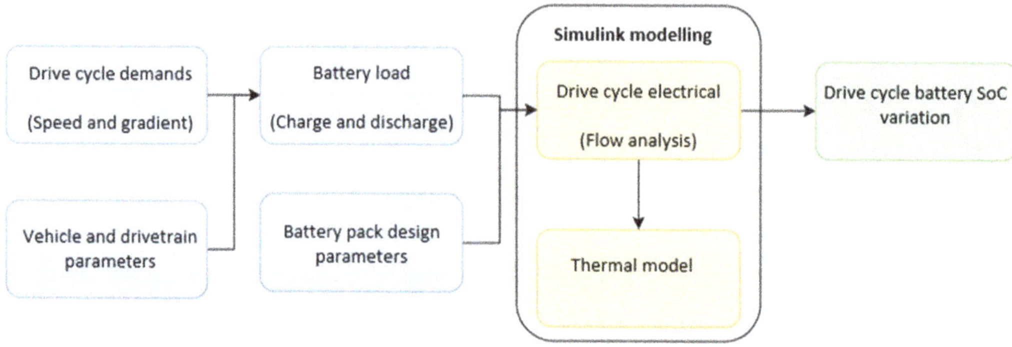

Figure 3. Electrical battery model design (inputs, modelling, and output).

During simulation, the cells' dynamic charge and discharge models were used to determine the cell terminal voltage based on current draw and time. Hence, the depth of discharge provided was defined by Equations (1) and (2), respectively. These models are pre-determined within MATLAB R2023b to suit a lithium-ion type battery, such as the cells used for this paper [46]. Table 4 shows the parameters used in the electrical model.

$$f_c(it,\ i*,\ i) = E_0 - K\left(\frac{Q}{it + 0.1Q}\right)i * -K\left(\frac{Q}{Q - it}\right)it + Ae^{-Bit} \tag{1}$$

$$f_d(it,\ i*,\ i) = E_0 - K\left(\frac{Q}{Q - it}\right)i * -K\left(\frac{Q}{Q - it}\right)it + Ae^{-Bit} \tag{2}$$

Table 4. Parameters of the electrical model.

Parameter	Symbol	Unit
Voltage	E_0	V
Polarisation constant	K	V/Ah
Current dynamics	$i*$	A
Battery current	i	A
Extracted capacity	it	Ah
Maximum battery capacity	Q	Ah
Exponential voltage	A	V
Exponential capacity	B	Ah^{-1}

2.2.2. Equivalent Circuit Model (ECM)

Methods of battery modelling can be split into two main categories, one being the electrochemical models (EM), and the second being the equivalent circuit model (ECM). The ECM utilises components like resistors, capacitors, and voltage sources to form a circuit network that replicates the behaviour and dynamics of a battery [47]. On the other hand, the EM provides a more precise depiction by accounting for the underlying electrochemical processes. The ECMs, particularly resistor–capacitor (RC) networks, are frequently used for lithium-ion battery modelling because they effectively capture the battery's dynamic behaviour.

Simulink's Simscape toolbox allowed the electrical system to be modelled using blocks representing various system components. The equivalent circuit model (ECM) implemented using the battery cell block in Simulink can be seen in Figure 4, alongside a schematic diagram.

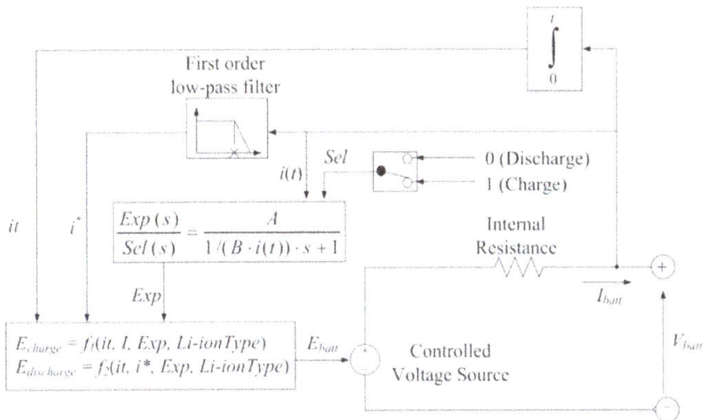

(a)

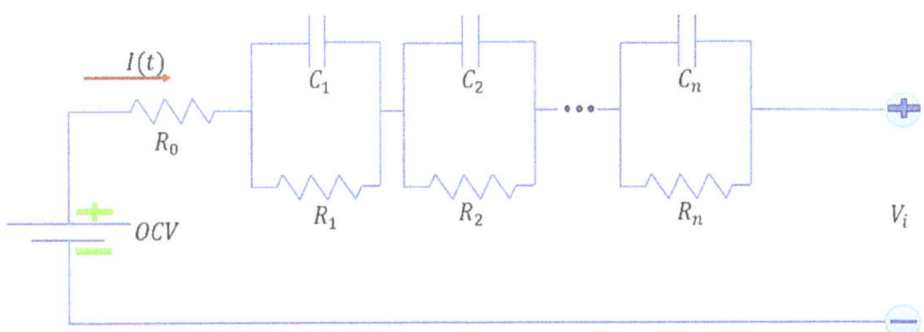

(b)

Figure 4. Schematic diagrams: (**a**) equivalent circuit model for Li-ion cell; (**b**) equivalent circuit model of nRC model.

Figure 4b illustrates a standard ECM configuration, specifically, the nRC model, which consists of an n-RC network. The battery's output voltage (V_i) depends on various parameters, such as the open circuit voltage (OCV), ohmic resistance (R_0), polarisation resistance (R_i), and the corresponding polarisation capacitance (C_i), as defined in Equation (3).

$$V_i = OCV - I_i\left[R_0 + \sum_{i=1}^{n} R_i\left(1 - e^{-\frac{t}{R_i C_i}}\right)\right] \tag{3}$$

2.2.3. Drive Cycle Energy Demand Analysis

This study considered two drive cycles—VECTO long haul and AVL These were chosen due to their use in the international emission assessment of modern trucks and their accurate representation of a typical heavy-duty truck drive cycle. The long-haul cycle predominantly features steady-state driving with some stop/start instances, similar to what could be expected of a drive along a motorway with moderate traffic. The AVL cycle is twice as long (200 km) and includes no stopping instances, with only some sections of reduced velocity. The AVL cycle, therefore, maintains a higher average velocity through the cycle and can be said to be representative of a drive through a mountainous region with fluctuations in altitude—the velocity and altitude profiles of each cycle can be seen in Figures 5 and 6.

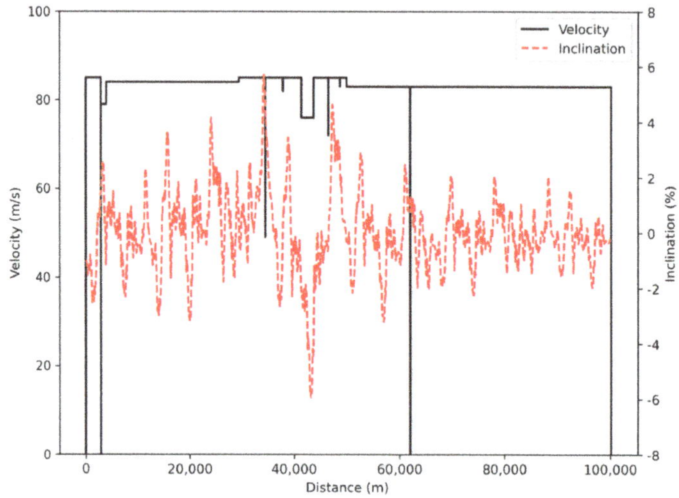

Figure 5. Velocity and inclination profiles for long haul.

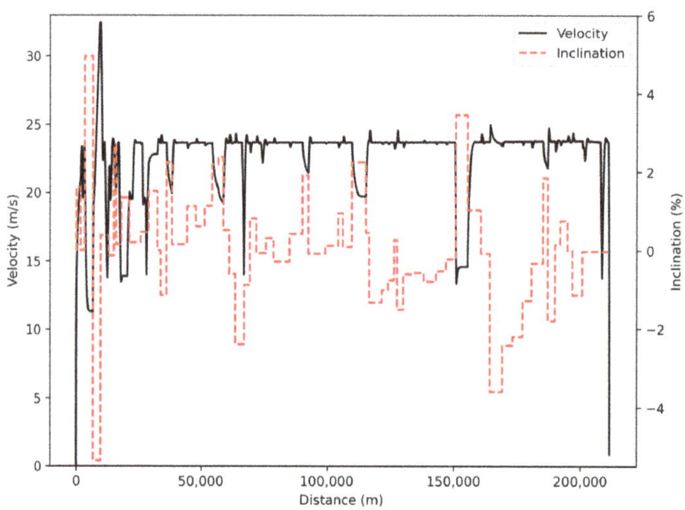

Figure 6. Velocity and inclination profiles for the AVL cycle.

2.3. Thermal Cooling Model

2.3.1. Thermal Parameters Overview

The battery pack materials used during the modelling process were selected based on the previously highlighted battery pack design procedure. The corresponding thermal parameters assumed for the materials were selected based on parameters used typically within existing papers [48–50] and, where available, manufacturer's datasheets. Table 5 provides an overview of the materials. The cooling system parameters were selected similarly, as shown in Table 6.

Table 5. Thermal parameters of the materials used for the battery pack [48–50].

Materials	Thickness (m)	Thermal Conductivity (W/(K∗m))	Application
Aluminium	0.004–0.005	202.40	Cell casing and module plates
HV Insulation	0.002	1.38	Electrical insulation and high-voltage protection
PVC Lid	0.005	0.02	Electrical insulation and thermal protection
Air Pockets	N/A	5.00	Heat Dissipation

Table 6. Cooling system parameters used in the current study [49,51–53].

System	Property	Values
Coolant	Viscosity	4.41 (mPa/s)
	Kinematic Viscosity	4.25 (mm^2/s)
	Thermal Conductivity	0.35 (W/mK)
	Specific Heat Capacity	3.57 (kJ/kg K)
Heat Exchanger Pipes	Thermal Conductivity	202.4 (W/mK)
	Hydraulic Diameter	0.01 (m)
	Pipe Roughness	1×10^{-6} (m)
	Laminar Flow Upper Reynolds Number	2300
	Turbulent Flow Lower Reynolds Number	4000
	Nusselt Number	3.66
	Darcy Friction Factor Constant	64

2.3.2. Simulink Model Overview

Simulink was again used as a simulation environment to model the battery pack and cooling system in 1D, as shown in Figures 7 and 8. The model components were connected electrically and thermally based on the arrangements in Table 5 to produce subsystems of each system level. Each component was assumed to be in contact with surrounding components, except at the pack level, where air spaces were assumed to be surrounding each pack. The battery subsystems were combined to create a representative model of the total battery, as shown in Figure 3.

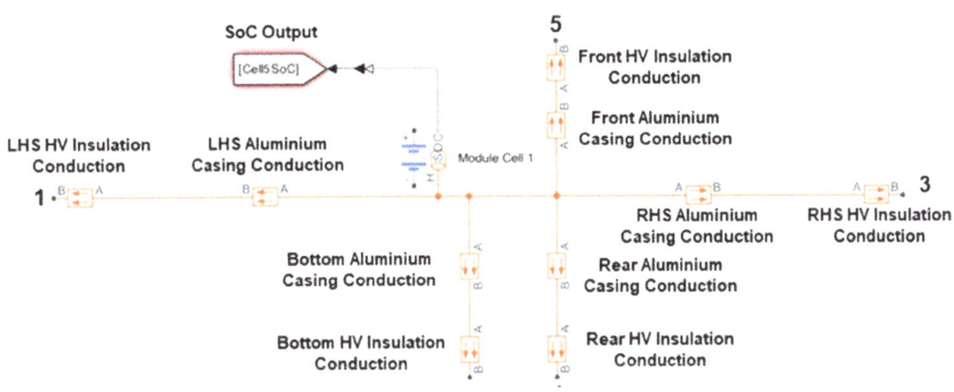

Figure 7. Battery cell thermal model in MATLAB/Simulink R2023b.

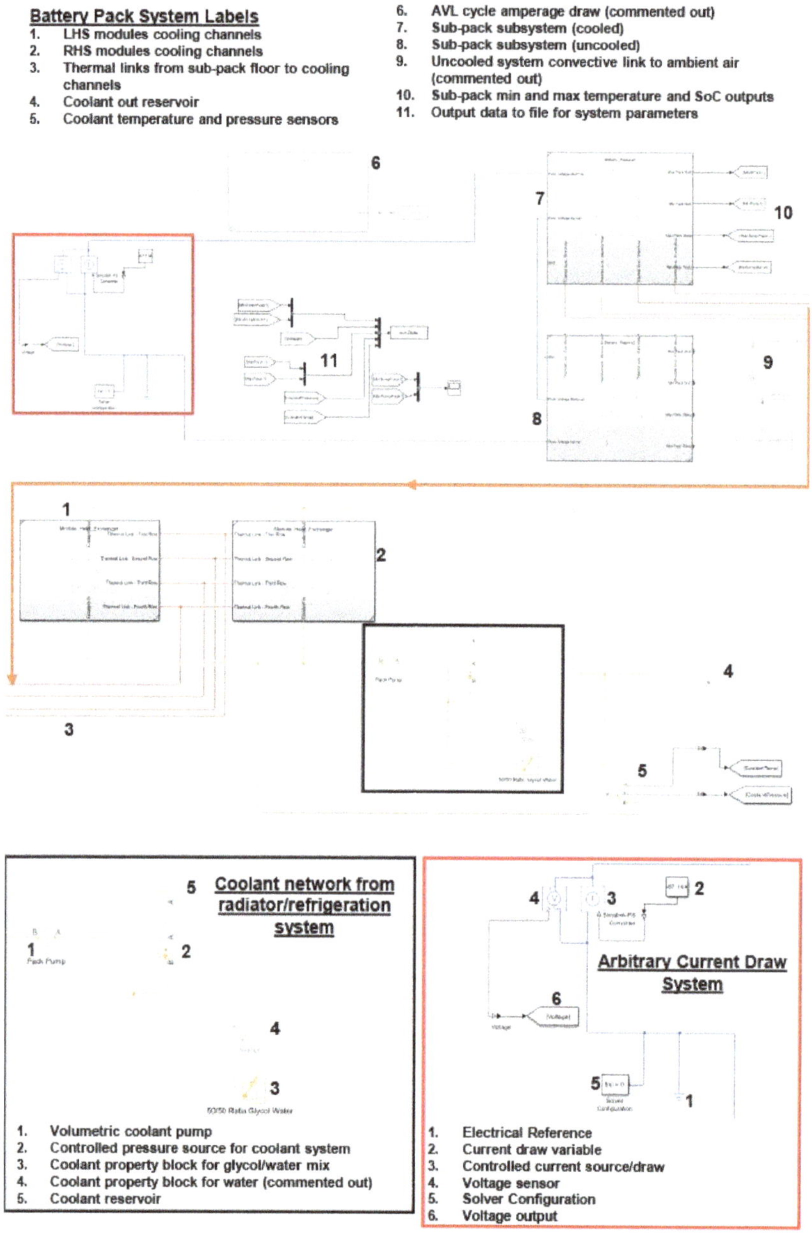

Figure 8. Overview of the 1D electro-thermal Simulink model used in the current study.

2.4. Simplified Assumptions for Battery System Modelling

The electrical battery model assumed uniform discharge, health degradation, and perfect cooling for all cells within the system. Therefore, the cell temperature was considered to be constant at 25 °C in line with the similar existing literature [32]. These assumptions meant that monitoring the electric state of a single cell was assumed sufficient and representative of all cells in the system.

All three models also used one fundamental assumption to reduce computational cost. It was assumed that one pack would provide a representative view of the whole system's behaviour. This was considered valid as the CAD model of the entire system suggested that each pack would be thermally isolated from each other with air. This was implemented by using one pack's cooling system and drawing a power demand for the pack with a scalar equal to $1/7^{th}$ of what would be faced by the total battery pack, therefore reducing computational cost by approximately 86%.

3. Results

3.1. Power Demand

The energy demand for a 44-tonne truck to complete both drive cycles was obtained from simulations carried out in AVL Cruise M, as shown in Figure 9. Both cycles included energy recovery from regenerative braking, as indicated by the positive power demand regions in both figures.

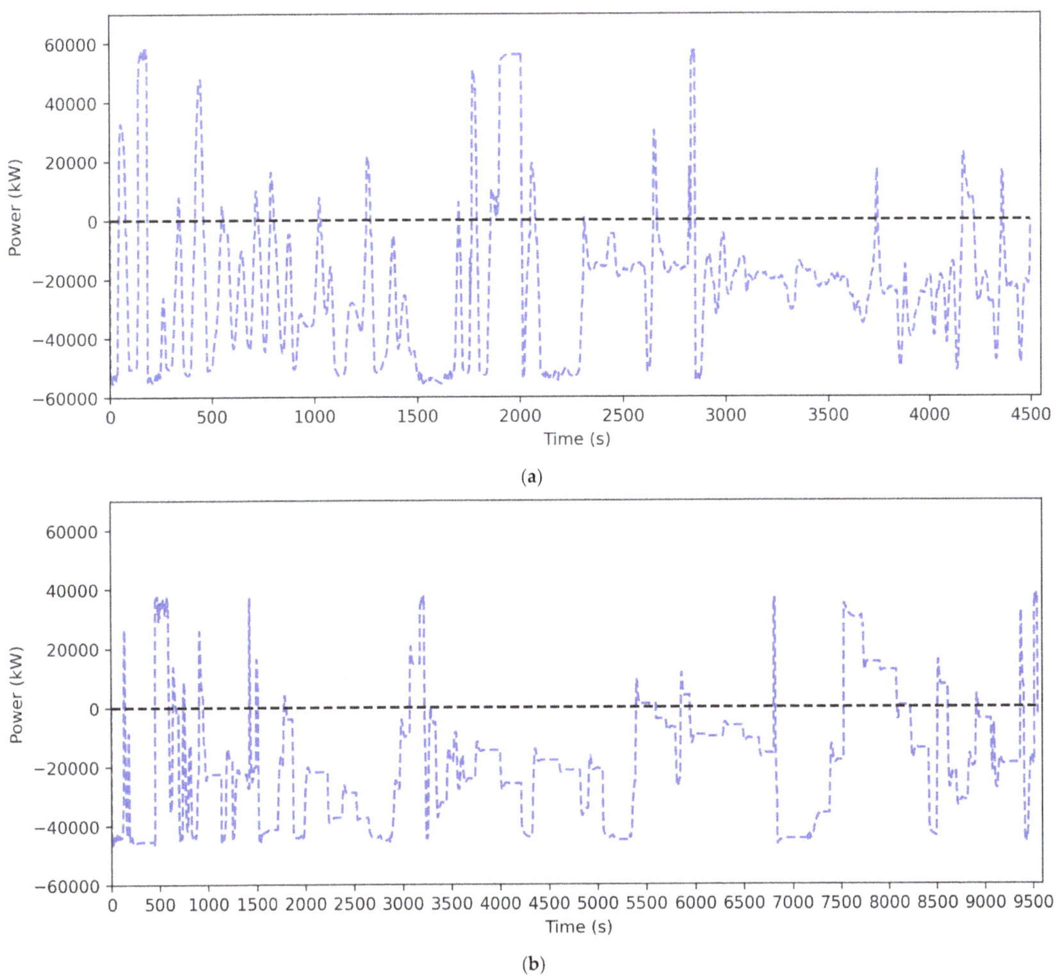

Figure 9. Power demand per pack for (**a**) long-haul cycle, (**b**) AVL cycle.

3.2. Electrical Battery Model

3.2.1. Model Validation

The electrical model was validated by assessing the total power consumption calculated from the initial and final SoC of the battery pack and scaling to consider the full, seven-pack system before comparing it to the literature [54,55]. For validation purposes, the SoH and initial SoC of the cells were taken as 100%. The more significant difference in energy consumption displayed by the long-haul cycle was attributed to the use of different cells in the literature and was therefore ignored. The results from the study can be seen in Table 7; from these, the model was deemed valid.

Table 7. Model simulation results.

Cycles	Energy Consumption—Literature (kWh)	Energy Consumption—Simulink Model (kWh)	Difference (%)
Long-haul	169.42	187.77	10.82
AVL	343.42	339.83	−1.05

3.2.2. Long-Haul Cycle

By varying the initial SoC and SoH values, nine use cases were obtained. The different use cases were investigated for energy consumption and post-cycle SoC, with the results shown below in Table 8 and Figure 10.

Table 8. Results of long-haul cycle.

Cycles	1			2000			4000		
Pack Energy Capacity (kWh)		66.41			59.78			53.13	
Initial SoC	1	0.9	0.8	1	0.9	0.8	1	0.9	0.8
Final SoC	0.60	0.48	0.37	0.55	0.44	0.32	0.48	0.37	0.26
Energy Consumed (kWh)	188	193	197	189	194	199	192	197	201

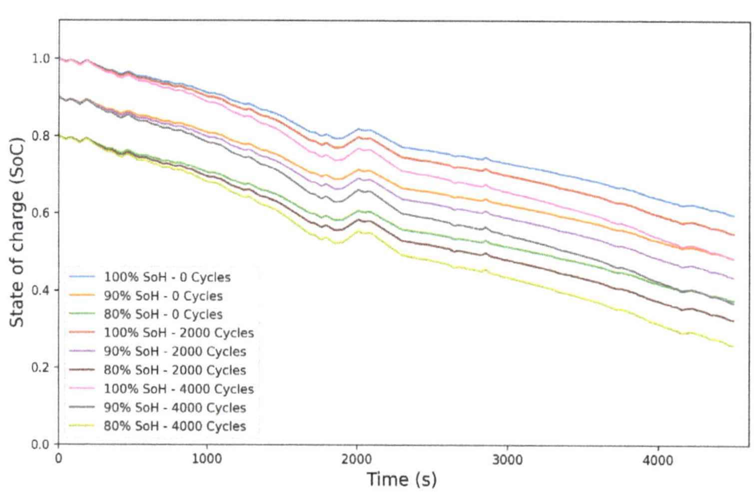

Figure 10. SoC profiles over time for different initial SoC and SoH conditions during long-haul cycle.

3.2.3. AVL Cycle Results

Equivalent use cases to those in Section 3.2.2 were used to analyse the AVL drive cycle for energy consumption and post-cycle SoC. The results are in Table 9 and Figure 11.

Table 9. Results of AVL cycle.

Cycles		1			2000			4000	
Pack Energy Capacity (kWh)		66.41			59.78			53.13	
Initial SoC	1	0.9	0.8	1	0.9	0.8	1	0.9	0.8
Final SoC	0.26	0.18	0.08	0.19	0.11	0	0.09	0	0
Energy Consumed (kWh)	342	333	335	339	331	335	335	335	298

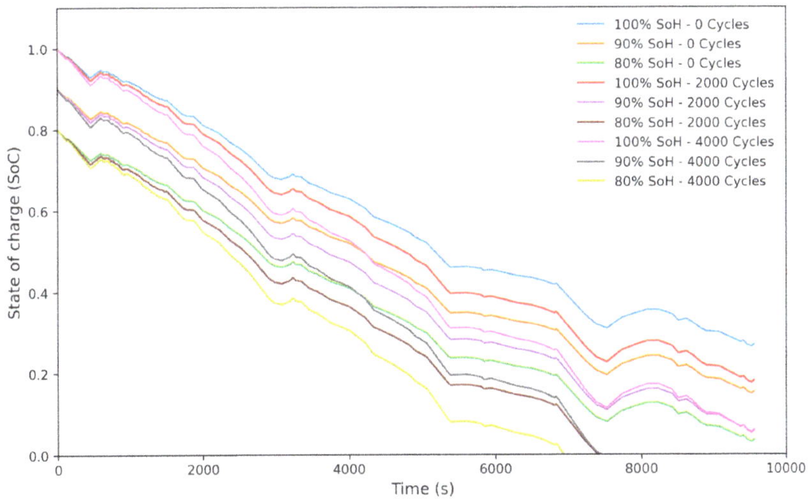

Figure 11. SoC profiles over time for different initial SoC and SoH conditions during AVL Cycle.

3.3. *Battery Thermal Management Performance*

3.3.1. Flow Rate Simulation Results

The thermal behaviour of the battery pack was simulated for 3600 s at the max C-rates obtained from the AVL cycle. These were 0.714 C and 0.935 C at 100 and 80% battery SoH, respectively, found during the electrical analysis. The median temperature increases and maximum temperature difference over the pack were recorded at varying coolant flow rates. Figures 12 and 13 show the effect of coolant flow rate on temperature increase and difference.

3.3.2. Inlet Simulation Results

The thermal response of the battery pack to varying inlet temperatures was also simulated for 3600 s at 0.714 C. Ambient temperatures of −15 °C, 25 °C and 40 °C were selected to cover both typical ambient temperatures and extremes that the literature has suggested heavy-duty trucks could face [29]. Figures 14 and 15 show how varying the inlet temperature affects battery temperature change.

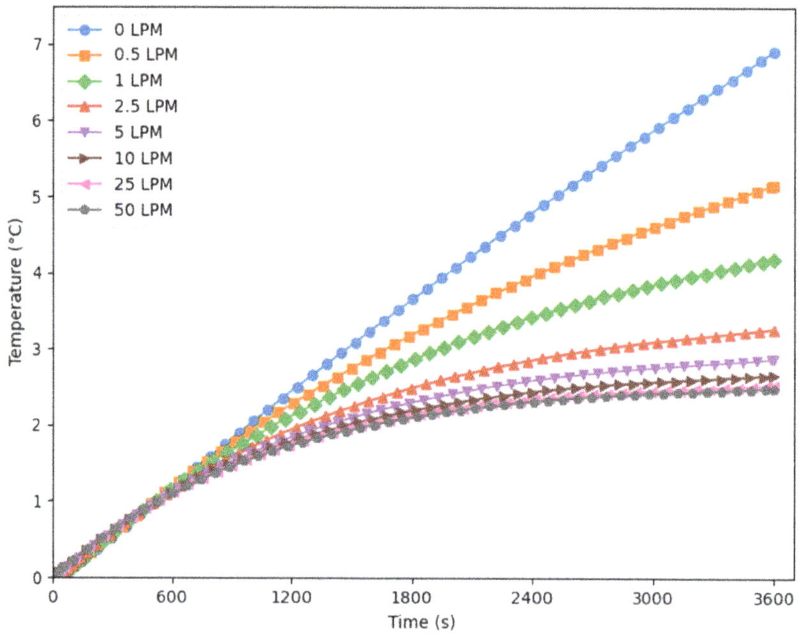

Figure 12. Median temperature increase in the battery pack exposed to 0.714 C at different coolant flow rates.

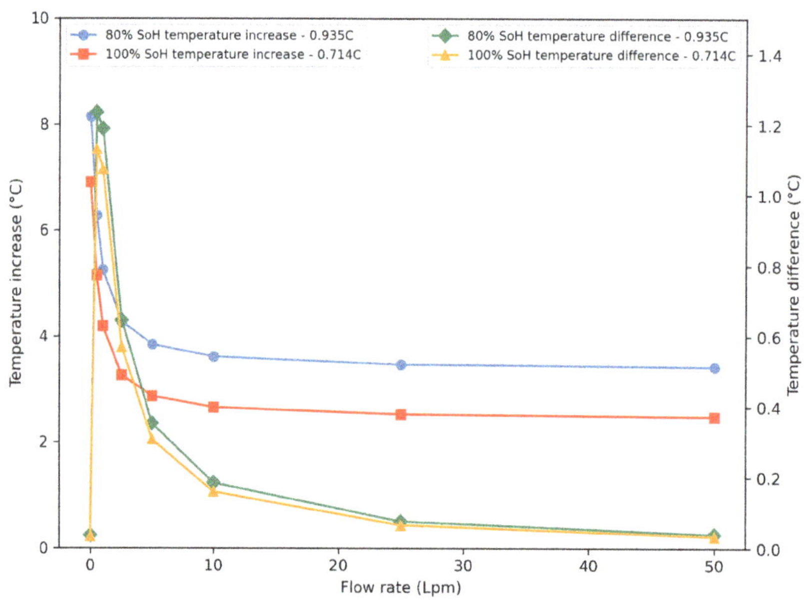

Figure 13. Temperature increase and difference at 3600 s of maximum C-rates from AVL drive cycle.

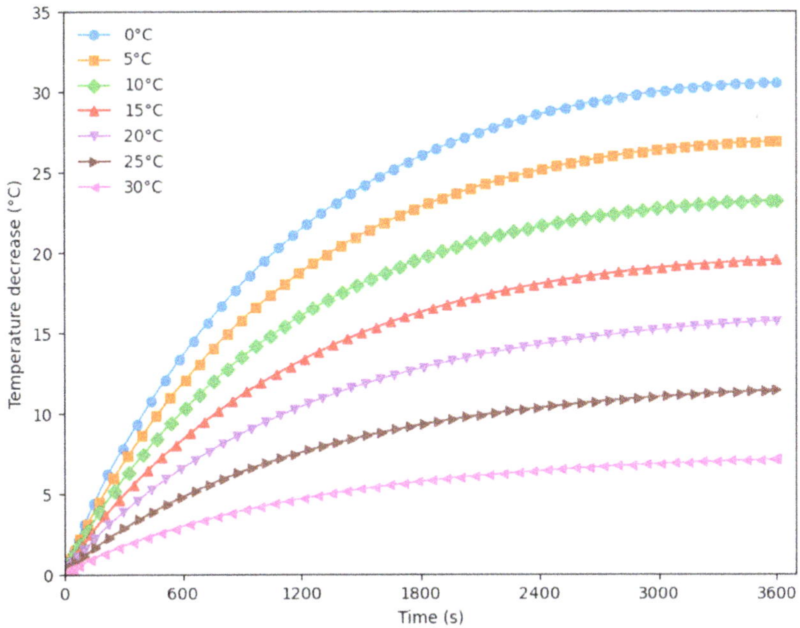

Figure 14. Decrease in temperature of battery pack at varying coolant inlet temperatures at 0.714 C (100% SoH) and 40 °C ambient temperature.

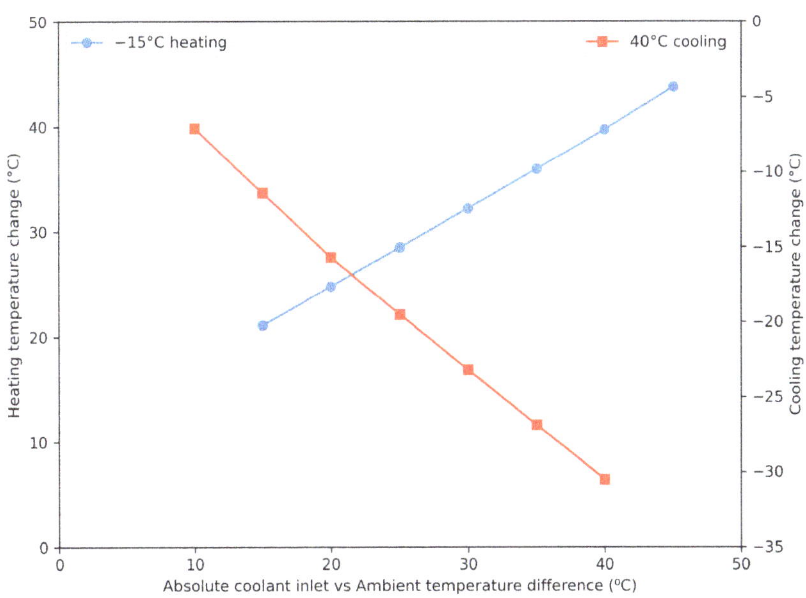

Figure 15. Temperature change in battery pack at varying coolant inlet temperatures at 3600 s of 0.714 C (100% SoH).

3.4. Transient Thermal Performance Analysis of the Battery Pack

Using the optimised BTMS parameters, the thermal behaviour of the truck's battery was analysed under AVL cycle loading at varying ambient temperatures. Each ambient temperature was studied without implementing a cooling system and with predefined optimal coolant parameters. These were determined from the results shown in Section 3.3 for each ambient temperature and can be seen in Table 10. Figure 16 shows the transient analysis results, highlighting the optimal region in grey.

Table 10. Optimised cooling system parameters.

Ambient Temperature (°C)	Coolant Flow Rate (LPM)	Coolant Inlet Temperature (°C)
−15	18	−7.2
25	18	25.0
40	18	32.2

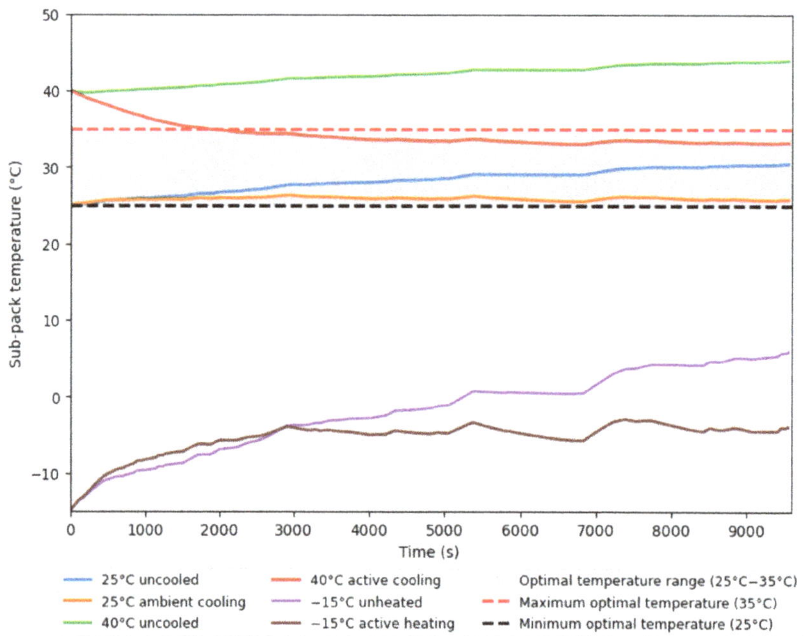

Figure 16. Battery pack temperature during AVL drive cycle loading.

4. Discussion

4.1. Performance Insights from the Electrical Model

4.1.1. Long-Haul Driving Cycle

Based on the post-cycle SoC of the pack, the total energy consumption was calculated to range between 188 and 201 kWh/100 km (~7% range) for the long-haul cycle, as shown in Table 8. This suggests a potential single-charge range of 247 km at full load under optimal operating conditions. As expected, energy consumption for the cycle increased with increased cycle ageing due to the increased internal resistance within each cell. Decreasing the initial SoC also increased energy usage due to the reduced pack voltage. In these scenarios, more current would be drawn to achieve the required power, compensating for the lower voltage. The use case with 100% SoH and SoC appeared to be the most energy efficient for the cycle, whilst expectedly, the oldest and least charged batteries appeared least efficient.

The final SoC values for the long-haul cycle, as shown in Figure 10, were found to vary from 60% to 26.7%, depending on the initial SoC and SoH. The SoC trend followed expectations from the power demand plot in Figure 9 and showed a steady discharge with one significant section of battery recharge around the 1750 s mark, coinciding with the large, positive peak in the power demand plot. It was also noticed that the power input in the recharge sectors resulted in a higher SoC increase for lower SoH cases. This occurred due to the reduced cell capacity, which allowed for a quicker charge at a constant power input. The impact of the remaining energy recovery sectors (around 200 s and 430 s) was found to be less significant, with the SoC increase not exceeding 1% for all cases. This likely reflects less extreme regenerative braking conditions in which gentle deceleration or less frequent breaking events occurred.

4.1.2. AVL Cycle

Unlike the long-haul cycle, the longer AVL cycle could not be completed in all nine use cases. Table 9 shows the total energy consumption varied between 331 and 342 kWh (3% range) when considering fully completed cycles, suggesting a potential battery range of 285 km under optimal conditions. However, as evident in Table 9, the 100% SoH, SoC battery was found to be the least efficient, contradicting the results from the long-haul cycle. This could be explained by the nature of the route, which contains more areas of regenerative braking, which is potentially more efficient at lower SoHs.

Final SoC values ranged from 26.4% and 0%, with three cases completely discharging the battery before the end of the cycle, as shown in Figure 11. Complete discharge was achieved within the 155 km–165 km range for these cases. The cycle contains three sectors of recharge, with the most significant one occurring around 7500 s. The minimum increase in SoC achieved at this part of the cycle was found to be around 4%. It can be estimated that without this large recharge power peak, a further three cases would not be able to complete the cycle.

4.1.3. Validation of Model

Overall, the model provided energy consumption results, which gave single-charge range predictions in line with current industry standards for electric class eight trucks [56]. The model was also validated by comparing energy consumption with 10.85% and −1.82% difference for the long-haul cycle and AVL cycle, respectively, upon comparison with the literature [56,57], and, hence, high confidence can be placed in the model. The energy flow in the system behaved as expected, showing the expected charge and discharge stages of the pack, accurately reflecting the power draw for each cycle, and giving reliable approximations for variations in SoC.

4.1.4. Key Assumptions and Limitations of the Model

The model utilised three main assumptions: constant temperature, uniform discharge, and uniform cycle ageing of all cells. The validity of these assumptions would be highly dependent on the performance of the cooling and battery management systems, both of which would be required for the battery design to be industrialised.

One modelling assumption that could have compromised the accuracy of results in comparison to experimental data was the neglect of the dynamic constants of the cells. The cells considered in this paper did not undergo characterisation tests, preventing accurate data regarding their behaviour under dynamic loading from being established. Only cycle ageing tests were carried out to understand the cycle ageing implications on the system performance in terms of vehicle single-charge driving range over the lifespan of the pack.

4.2. Thermal Cooling System Model

4.2.1. Coolant Flow Rate

Figures 12 and 13 suggest that with an ambient inlet temperature, cooling performance increases with coolant flow rate, up to a point of diminishing returns. For example, an increase in flow rate from 1 to 10 LPM at 0.714 C decreased the final temperature increase by 36.5% and temperature difference by 85.1%. The increase in cooling performance is likely due to the greater flow rate providing increased cooling capacity to the system, with more mass able to absorb heat each second, therefore maintaining the temperature gradient. However, increasing the coolant flow rate from 10 to 25 LPM saw limited additional cooling benefits, suggesting that the rate of heat transfer was no longer limited by the flow rate.

Additionally, Figure 13 suggests that greater C-rates increase temperature difference across a battery pack, with this decreasing slightly with an increase in flow rate. This is likely due to the increase in heat production and, therefore, overall pack temperatures. In turn, this would likely cause greater heat transfer to the coolant across the pack, reducing the temperature gradient at the coolant outlet, and thus reducing the rate of heat transfer from the end modules compared to those at the inlet.

4.2.2. Coolant Inlet Temperature

Figure 15 shows a positive relationship between coolant inlet temperature and thermal benefits, with correlation coefficients of 0.776 for cooling and 0.753 for heating obtained using the least-square method. Increased modification of the coolant temperature at extreme ambient temperatures would likely provide continued thermal benefits. However, due to the power demand associated with active coolant temperature modification, the BTMS should be optimised for each ambient temperature to minimise parasitic energy consumption. This can be achieved by implementing dynamic control of the thermal management system [57].

4.2.3. System Optimisation

Results shown in Section 3.3 and the CFD results [58] were used to optimise the BTMS. Figure 13 suggested that 18 LPM would be the optimal coolant flow rate per sub-pack, with this value found to be achievable in typical automotive applications [44]. This was determined by combining the point of diminishing returns, ~15 LPM, with a SF (safety factor) of 1.2. SFs of 1.2–1.5 have been suggested for battery systems [59], with 1.2 selected to minimise parasitic power consumption seen with increasing flow rates [37]. CFD results in [58] showed that at optimal ambient temperatures, ambient cooling would suffice. At extremes, it was found that an active system would be required, with 7.8 °C of active cooling or heating to be optimal [60].

4.3. Combined Transient Analysis Results

Figure 16 shows that the optimised cooling system demonstrated effective performance under ambient temperature conditions of 25 °C and 40 °C. When applied throughout the AVL drive cycle, the system enhanced the battery's thermal stability by 87.7% at 25 °C, ensuring the battery remained in the optimal temperature range for the entire drive cycle and 81% of the cycle at 40 °C.

This likely means that at ambient temperatures of 25 °C and above, the truck's battery would be thermally stable with the suggested parameters, avoiding issues like runaway, whilst achieving optimal performance on typical drive cycles.

Despite optimisation, Figure 16 showed poor BTMS performance at sub-zero temperatures, and the system's efficiency was not evaluated. These could likely be improved through the implementation of a battery preheating controller to improve low-temperature performance, with studies showing 40 °C temperature rises within an hour being possible [60]. This would allow the truck's battery to operate in its optimal temperature range with just 1 h of preheating. Studies have also shown that P.I.D style controllers can allow for reductions in operational time of BTMS during driving cycles, whilst maintaining thermal stability [61]. These could be used to reduce parasitic power draw, ultimately helping to improve battery range and BTMS efficiency.

5. Conclusions

This study develops a novel methodology, presenting the findings of an investigation into the electrical and thermal performance of a battery pack designed for a heavy-duty truck. The analysis employed 1D simulation techniques, utilising AVL CRUISE M and MATLAB Simulink Simscape R2023b, to evaluate the system under representative dynamic driving conditions. This approach allowed for a detailed assessment of energy consumption, state of charge (SoC), temperature fluctuations, and the effects of various cooling configurations while minimising computational costs. The simulations were conducted over two real-world drive cycles, providing valuable insights into the performance of the proposed battery pack design. The key findings of this investigation are as follows:

- As cell ageing progresses and the initial SoC decreases, the suitability of the battery pack for long-distance travel diminishes significantly, indicating that battery degradation plays a critical role in reducing range capabilities.
- Despite ageing, cells may exhibit improved range performance from regenerative braking during drive cycles, suggesting that drive cycle regeneration can partially offset the impacts of cell degradation.
- Increasing the coolant flow rate to 18 LPM and actively controlling the inlet temperature within ±7.8 °C were found to significantly enhance thermal regulation, resulting in an 80%+ improvement in thermal stability under ambient conditions of 25 °C and 40 °C.
- The predicted single-charge range of the vehicle varied based on drive conditions, with a maximum range of 285 km, which underscores the variability in battery performance depending on real-world operating conditions.

The integration of AVL CRUISE M and MATLAB/Simulink Simscape R2023b in the simulations provided a comprehensive platform for both electrical and thermal analysis, offering a high degree of accuracy and flexibility. These tools were crucial in capturing the complex interactions between the battery pack's electrical behaviour and its thermal management system.

While this study's results offer a solid foundation for advancing electric propulsion systems in heavy-duty trucks, further design optimisation and experimental validation are needed. Future research should focus on validating the electrical and thermal performance of the proposed system through practical testing, incorporating parasitic power draws from auxiliary components such as pumps and compressors into the simulations, considering the effects of charging system behaviour, and developing a more robust BTMS tailored for electric heavy-duty vehicles. Also, higher C-rates should be considered for the analysis.

This investigation provides valuable insights into the performance and feasibility of battery packs in heavy-duty electric trucks, especially concerning their thermal and electrical characteristics in dynamic driving conditions. This study highlights the need for further cooling optimisation and increased battery longevity. As electric mobility continues to evolve, this research will contribute to the goal of achieving more sustainable and efficient heavy-duty electric vehicles.

Author Contributions: M.M.: Conceptualisation, Methodology, Investigation, Formal Analysis, Validation, Software, Writing—Original Draft, and Data Curation; T.S.: Conceptualisation, Methodology, Investigation, Formal Analysis, Validation, Software, Writing—Original Draft, and Data Curation; U.T.: Investigation, Formal Analysis, Software, Writing—Original Draft, and Data Curation; F.S.: Conceptualisation, Methodology, Supervision, and Writing—Review and Editing; H.T.: Writing—Review and Editing, and Supervision; M.B.: Conceptualisation, Methodology, Software, Supervision, and Writing—Review and Editing. All authors have read and agreed to the published version of the manuscript.

Funding: This research received no external funding.

Data Availability Statement: The original data presented in the study are included in the article, and further inquiries can be directed to the corresponding author.

Acknowledgments: The authors acknowledge AVL List GmbH's support for providing the simulation tools for the University of Leeds through their University Partnership Programme is appreciated.

Conflicts of Interest: Author Farhad Salek was employed by the company AVL Powertrain UK Limited. The remaining authors declare that the research was conducted in the absence of any commercial or financial relationships that could be construed as a potential conflict of interest.

References

1. Hansen, J.; Ruedy, R.; Sato, M.; Lo, K. Global surface temperature change. *Rev. Geophys.* **2010**, *48*, RG000345. [CrossRef]
2. Hannah, R. Cars, Planes, Trains: Where Do CO_2 Emissions from Transport Come from? Our World in Data. 2020. Available online: https://ourworldindata.org/co2-emissions-from-transport (accessed on 14 April 2024).
3. European Parliament. Regulation (EU) 2019/631 of the European Parliament and of the Council of 17 April 2019 setting CO_2 emission performance standards for new passenger cars and for new light commercial vehicles, and repealing Regulations (EC) No 443/2009 and (EU) No 510/2011. *Off. J. Eur. Union* **2021**, *L 111*, 13–53.
4. Department of Transport, Government Takes Historic Step Towards Net-Zero with End of Sale of New Petrol and Die-Sel Cars by 2030. 2020. Available online: https://www.gov.uk/government/news/government-takes-historic-step-towards-net-zero-with-end-of-sale-of-new-petrol-and-diesel-cars-by-2030 (accessed on 15 April 2024).
5. Abnett, K. Electric Car Sales Surge as Europe's Climate Targets Bite. 2021. Available online: https://www.reuters.com/business/sustainable-business/electric-car-sales-surge-europes-climate-targets-bite-2021-06-29/ (accessed on 16 April 2024).
6. Thomas, C.E.S. Transportation options in a carbon-constrained world: Hybrids, plug-in hybrids, biofuels, fuel cell electric vehicles, and battery electric vehicles. *Int. J. Hydrogen Energy* **2009**, *34*, 9279–9296. [CrossRef]
7. Department for Transport. UK Confirms Pledge for Zero-Emission HGVs by 2040 and Unveils New Chargepoint Design. 2021. Available online: https://www.gov.uk/government/news/uk-confirms-pledge-for-zero-emission-hgvs-by-2040-and-unveils-new-chargepoint-design#:~:text=News%20story-,UK%20confirms%20pledge%20for%20zero-emission%20HGVs%20by%202040%20and,within%20the%20next%202%20decades.&text=All%20new%20heavy%20goods%20vehicles,today%20(10%20November%202021) (accessed on 20 November 2024).
8. Hao, H.; Geng, Y.; Tate, J.E.; Liu, F.; Chen, K.; Sun, X.; Liu, Z.; Zhao, F. Impact of transport electrification on critical metal sustainability with a focus on the heavy-duty segment. *Nat. Commun.* **2019**, *10*, 5398. [CrossRef]
9. Agency, I.E. Global EV Outlook 2021: Accelerating Ambitions Despite the Pandemic. 2021. Available online: www.iea.org/t&c/ (accessed on 15 April 2024).
10. Transport, D.O. A Simplified Guide to Lorry Types and Weights. 2023. Available online: https://assets.publishing.service.gov.uk/media/5a74dbd340f0b65f61322ceb/simplified-guide-to-lorry-types-and-weights.pdf (accessed on 16 April 2024).
11. Hasnain, A.; Gamwari, A.S.; Resalayyan, R.; Sadabadi, K.F.; Khaligh, A. Medium and heavy duty vehicle electrification: Trends and technologies. *IEEE Trans. Transp. Electrif.* **2024**. [CrossRef]
12. Sandeep, V.; Shastri, S.; Sardar, A.; Salkuti, S.R. Modeling of battery pack sizing for electric vehicles. *Int. J. Power Electron. Drive Syst.* **2020**, *11*, 1987–1994. [CrossRef]
13. Salek, F.; Morrey, D.; Henshall, P.; Resalati, S. *Techno-Economic Assessment of Utilising Second-Life Batteries in Electric Vehicle Charging Stations*; SAE: Warrendale, PA, USA, 2023. [CrossRef]

14. Tudoroiu, R.E.; Zaheeruddin, M.; Tudoroiu, N.; Radu, S.M. SOC estimation of a rechargeable li-ion battery used in fuel cell hybrid electric vehicles—Comparative study of accuracy and robustness performance based on statistical criteria. Part II: SOC estimators. *Batteries* **2020**, *6*, 41. [CrossRef]
15. Basma, H.; Beys, Y.; Rodríguez, F. Battery Electric Tractor-Trailers in the European Union: A Vehicle Technology Analysis. 2021. Available online: www.theicct.org (accessed on 15 April 2024).
16. Yao, L.W.; Aziz, J.A.; Kong, P.Y.; Idris, N.R.N. Modeling of lithium-ion battery using MATLAB/simulink. In Proceedings of the IECON 2013—39th Annual Conference of the IEEE Industrial Electronics Society, Vienna, Austria, 10–13 November 2013; pp. 1729–1734.
17. Salek, F.; Azizi, A.; Resalati, S.; Henshall, P.; Morrey, D. Mathematical modelling and simulation of second life battery pack with heterogeneous state of health. *Mathematics* **2022**, *10*, 3843. [CrossRef]
18. Martyushev, N.V.; Malozyomov, B.V.; Sorokova, S.N.; Efremenkov, E.A.; Qi, M. Mathematical modeling of the state of the battery of cargo electric vehicles. *Mathematics* **2023**, *11*, 536. [CrossRef]
19. Pesaran, A.A. Battery thermal models for hybrid vehicle simulations. *J. Power Sources* **2002**, *110*, 370–382. [CrossRef]
20. Pesaran, A.; Santhanagopalan, S.; Kim, G.H. *Addressing the Impact of Temperature Extremes on Large Format Li-Ion Batteries for Vehicle Applications (Presentation)*; National Renewable Energy Lab.(NREL): Golden, CO, USA, 2013.
21. Tarascon, J.-M.; Gozdz, A.S.; Schmutz, C.; Shokoohi, F.; Warren, P.C. Performance of Bellcore's plastic rechargeable Li-ion batteries. *Solid State Ion.* **1996**, *86*, 49–54. [CrossRef]
22. Lo, J. Effect of Temperature on Lithium-Iron Phosphate Battery Performance and Plug-In Hybrid Electric Vehicle Range. Master of Applied Science Thesis, University of Waterloo, Waterloo, ON, USA, 2013.
23. Kim, J.; Oh, J.; Lee, H. Review on battery thermal management system for electric vehicles. *Appl. Therm. Eng.* **2019**, *149*, 192–212. [CrossRef]
24. Choudhari, V.G.; Dhoble, D.A.S.; Sathe, T.M. A review on effect of heat generation and various thermal management systems for lithium ion battery used for electric vehicle. *J. Energy Storage* **2020**, *32*, 101729. [CrossRef]
25. Greco, A.; Cao, D.; Jiang, X.; Yang, H. A theoretical and computational study of lithium-ion battery thermal management for electric vehicles using heat pipes. *J. Power Sources* **2014**, *257*, 344–355. [CrossRef]
26. Saw, L.H.; Ye, Y.; Tay, A.A.O.; Chong, W.T.; Kuan, S.H.; Yew, M.C. Computational fluid dynamic and thermal analysis of Lithium-ion battery pack with air cooling. *Appl. Energy* **2016**, *177*, 783–792. [CrossRef]
27. Tesla, Dimensions and Weights. 2024. Available online: https://www.tesla.com/ownersmanual/model3/en_cn/GUID-56562137-FC31-4110-A13C-9A9FC6657BF0.html (accessed on 13 November 2024).
28. Nissan. Leaf, the Electric Family Car. 2024. Available online: https://www.nissan.co.uk/vehicles/new-vehicles/leaf.html?cid=psm_cmid=17064511648_grid=131231644970_adid=595105867889&gad_source=1&gclid=CjwKCAiAudG5BhAREiwAWMlSjF4SKaC72zA2ccqbWAoHdD7rTwQSaUZfqC8-JupvA95B2gdF-CqO1BoCtvAQAvD_BwE&gclsrc=aw.ds (accessed on 13 November 2024).
29. Volvo. FH Specifications. 2024. Available online: https://www.volvotrucks.co.uk/en-gb/trucks/models/volvo-fh/specifications.html (accessed on 13 November 2024).
30. Gunes, F.; Yang, Y.; Kim, B.; Wellers, M. Mechanical hybrid control for heavy-goods vehicle. In Proceedings of the 2018 UKACC 12th International Conference on Control, Sheffield, UK, 5–7 September 2018; pp. 127–132. [CrossRef]
31. Singh, A.K.; Dalai, A.; Kumar, P. Analysis of induction motor for electric vehicle application based on drive cycle analysis. In Proceedings of the 2014 IEEE International Conference on Power Electronics, Drives and Energy Systems, Mumbai, India, 16–19 December 2014; pp. 1–6. [CrossRef]
32. Salameh, M.; Brown, I.P.; Krishnamurthy, M. Fundamental evaluation of data clustering approaches for driving cycle-based machine design optimization. *IEEE Trans. Transp. Electrif.* **2019**, *5*, 1395–1405. [CrossRef]
33. Leonard, A.T.; Salek, F.; Azizi, A.; Resalati, S. Electrification of a class 8 heavy-duty truck considering battery pack sizing and cargo capacity. *Appl. Sci.* **2022**, *12*, 9683. [CrossRef]
34. Pesaran, A. Battery thermal management in EVs and HEVs: Issues and solutions. In Proceedings of the Advanced Automotive Battery Conference, Las Vegas, NV, USA, 6–8 February 2001; p. 10.
35. Xu, X.; Li, W.; Xu, B.; Qin, J. Numerical study on a water cooling system for prismatic LiFePO4 batteries at abused operating conditions. *Appl. Energy* **2019**, *250*, 404–412. [CrossRef]
36. Piao, C.; Chen, T.; Zhou, A.; Wang, P.; Chen, J. Research on electric vehicle cooling system based on active and passive liquid cooling. *J. Phys. Conf. Ser.* **2020**, *1549*, 042146. [CrossRef]
37. Lan, C.; Xu, J.; Qiao, Y.; Ma, Y. Thermal management for high power lithium-ion battery by minichannel aluminum tubes. *Appl. Therm. Eng.* **2016**, *101*, 284–292. [CrossRef]
38. Karimi, D.; Behi, H.; Hosen, M.S.; Jaguemont, J.; Berecibar, M.; Van Mierlo, J. A compact and optimized liquid-cooled thermal management system for high power lithium-ion capacitors. *Appl. Therm. Eng.* **2020**, *185*, 116449. [CrossRef]

39. Yue, Q.L.; He, C.X.; Wu, M.C.; Xu, J.B.; Zhao, T.S. Pack-level modeling of a liquid cooling system for power batteries in electric vehicles. *Int. J. Heat Mass Transf.* **2022**, *192*, 122946. [CrossRef]
40. Broatch, A.; Olmeda, P.; Margot, X.; Agizza, L. A generalized methodology for lithium-ion cells characterization and lumped electro-thermal modelling. *Appl. Therm. Eng.* **2022**, *217*, 119174. [CrossRef]
41. Samsung. *Introduction of Samsung SDI's 94 Ah Cells.* 2015. Available online: https://files.gwl.eu/inc/_doc/attach/StoItem/7213/30118_Introduction%20of%20SDI%20EV%2094Ah%20cell_V9-2.pdf (accessed on 12 November 2024).
42. Wang, Q.; Jiang, B.; Li, B.; Yan, Y. A critical review of thermal management models and solutions of lithium-ion batteries for the development of pure electric vehicles. *Renew. Sustain. Energy Rev.* **2016**, *64*, 106–128. [CrossRef]
43. Boyd, J. China and Japan push for a global charging standard for EVs. *IEEE Spectr.* **2018**, *56*, 12–13. [CrossRef]
44. Vovyo Pump, Automotive Electric Water Pump. 2024. Available online: https://www.vovyopump.com/automotive-electric-water-pump/ (accessed on 15 April 2024).
45. Hoşöz, M.; Gündem, A.; Keklik, E. Performance comparison of propylene glycol-water and ethylene glycol-water mixtures as engine coolants in a flat-tube automobile radiator. *Int. J. Automot. Sci. Technol.* **2021**, *5*, 147–156. [CrossRef]
46. Mathworks. Battery: Generic Battery Model; n.d. Available online: https://uk.mathworks.com/help/sps/powersys/ref/battery.html (accessed on 12 November 2024).
47. Lai, X.; Gao, W.; Zheng, Y.; Ouyang, M.; Li, J.; Han, X.; Zhou, L. A comparative study of global optimization methods for parameter identification of different equivalent circuit models for Li-ion batteries. *Electrochim. Acta* **2019**, *295*, 1057–1066. [CrossRef]
48. Zhang, A.; Li, Y. Thermal conductivity of aluminum alloys—A review. *Materials* **2023**, *16*, 2972. [CrossRef]
49. ANSYS Inc. *ANSYS Fluent Theory Guide*; ANSYS Inc.: Canonsburg, PA, USA, 2023.
50. Mazhar, S.; Qarni, A.A.; Haq, Y.U.; Haq, Z.U.; Murtaza, I. Promising PVC/MXene based flexible thin film nanocomposites with excellent dielectric, thermal and mechanical properties. *Ceram. Int.* **2020**, *46*, 12593–12605. [CrossRef]
51. Dow Chemical Company. Engineering and Operating Guide for Dowfrost and Dowfrost HD inhibited Propylene Glycol-Based Heat Transfer Fluids. 2008. Available online: https://www.dow.com/documents/180/180-01286-01-engineering-and-operating-guide-for-dowfrost-and-dowfrost-hd.pdf?iframe=true#:~:text=Composition%20(%25%20by%20weight)%20Propylene%20Glycol%2095.5%2094.0,10.7%20Reserve%20Alkalinity%20(min.)%209.0%20ml%2016.0%20ml (accessed on 12 November 2024).
52. MathWorks. MATLAB Simulink. 2023. Available online: https://www.mathworks.com/ (accessed on 4 December 2024).
53. Durst, F.; Arnold, I. *Fluid Mechanics: An Introduction to the Theory of Fluid Flows*; Springer: Berlin/Heidelberg, Germany, 2008.
54. Salek, F.; Halder, P.; Leonard, A.T.; Babaie, M.; Resalati, S.; Zare, A. *Battery Sizing, Parametric Analysis, and Powertrain Design for a Class 8 Heavy-Duty BATTERY Electric Truck*; SAE Technical Paper Series; SAE: Warrendale, PA, USA, 2023. Available online: https://api.semanticscholar.org/CorpusID:258034060 (accessed on 12 November 2024).
55. Salek, F.; Abouelkhair, E.; Babaie, M.; Cunliffe, F.; Zare, A. Assessment of the powertrain electrification for a heavy-duty class 8 truck for two different electric drives. In Proceedings of the SAE Powertrains, Fuels & Lubricants Conference & Exhibition, Krakow, Poland, 6–8 September 2022; p. 8.
56. DAF. New Generation DAF Electric. 2024. Available online: https://www.daf.co.uk/en-gb/trucks/new-generation-daf-electric (accessed on 17 November 2024).
57. Kim, E.; Shin, K.G.; Lee, J. Real-time battery thermal management for electric vehicles. In Proceedings of the 2014 ACM/IEEE International Conference on Cyber-Physical Systems (ICCPS), Berlin, Germany, 14–17 April 2014; pp. 72–83. [CrossRef]
58. Makings, M.; Maciocha, M.; Biggs, J.; Salek, F.; Zare, A.; Resalati, S.; Short, T.; Babaie, M. *CFD Analysis of the Battery Thermal Management System for a Heavy-Duty Truck*; WCX SAE World Congress Experience; SAE Technical Paper; SAE: Warrendale, PA, USA, 2024. [CrossRef]
59. Henke, M.; Hailu, G. Thermal management of stationary battery systems: A literature review. *Energies* **2020**, *13*, 4194. [CrossRef]
60. Wu, S.; Xiong, R.; Li, H.; Nian, V.; Ma, S. The state of the art on preheating lithium-ion batteries in cold weather. *J. Energy Storage* **2020**, *27*, 101059. [CrossRef]
61. Angermeier, S.; Ketterer, J.; Karcher, C. Liquid-based battery temperature control of electric buses. *Energies* **2020**, *13*, 4990. [CrossRef]

batteries

MDPI

Article

New BMS Topology with Active Cell Balancing Between Electric Vehicles' Traction and Auxiliary Batteries

José Gabriel O. Pinto [1,*], Manuel Freitas Silva [1], Luis A. M. Barros [1] and José A. Afonso [2,*]

[1] ALGORITMI Research Centre/LASI, University of Minho, 4800-058 Guimarães, Portugal; a54377@alumni.uminho.pt (M.F.S.); lbarros@dei.uminho.pt (L.A.M.B.)
[2] Center for Microelectromechanical Systems (CMEMS), University of Minho, 4800-058 Guimarães, Portugal
* Correspondence: gpinto@dei.uminho.pt (J.G.O.P.); jose.afonso@dei.uminho.pt (J.A.A.)

Abstract: This paper proposes a new topology for a battery management system (BMS) with active cell balancing capable of exchanging energy between an electric vehicle's traction and auxiliary batteries. This topology facilitates energy exchange between any cell in the traction battery pack and with the auxiliary battery. The proposed topology allows both the selection of the cells involved in the balancing process and the charging of the auxiliary battery, eliminating the need for a dedicated dc-dc isolated power converter. The flexibility of this topology allows the adoption of different balancing strategies, which can be used to improve balancing efficiency. The proposed topology was first analyzed through computer simulations, and a laboratory BMS prototype was developed. The results from the simulation and experimental tests validate the topology operation and its performance in transferring energy between the cells and the auxiliary battery.

Keywords: active cell balancing; battery; battery management system; electric vehicles; electric mobility

1. Introduction

Electric vehicles (EVs) are considered one of the leading technologies needed to reduce the environmental impact of the transportation sector [1]. Nowadays, several fully electric vehicles are under development, from light vehicles to cars and heavy vehicles, with several hundred models already commercially available [2]. Regarding aircraft, electrification is at an earlier stage, but it is expected to follow the same trend as land vehicles [3].

One of the most critical EV components is the energy storage and management system, which requires high capacity, low weight, compactness, and effective thermal management [4]. EV energy is generally stored in a battery pack containing several elementary cells connected in and/or parallel. To ensure that the cells operate safely and efficiently, they are usually equipped with a supervisory electronic circuit known as a battery management system (BMS) [5]. A BMS should perform monitoring and protection functions, contributing to the integrity and longevity of the cells. To achieve this, it is necessary to measure parameters, such as voltage, current, and operating temperature, to estimate parameters, such as the state of charge (SoC), depth of discharge (DoD), state of health (SoH), and internal resistance [6].

The determination of the SoC is one of the key challenges when dealing with batteries, as discussed in [7]. This is because the methods available for estimating the SoC only approximate this value. The accuracy of SoC estimation can vary significantly depending on the type of battery and the available processing power. The most common methods for estimating the SoC are [8]: open circuit voltage-based methods (OCVMs); Coulomb counting (Ah Counting); resistance spectroscopy-based methods (ISBMs); model-based methods (MBMs); and artificial neural network-based methods (ANNBMs). OCVMs are simple to apply, relying on a predefined relationship between the open circuit voltage (OCV) and the SoC. However, during the operation of most batteries, this relationship can change with temperature and battery deterioration, making the estimation more challenging. To obtain more accurate results, in the case of lithium cells, as described in [8], the battery must be disconnected from the circuit and left to rest during its relaxation time, enabling the OCV to stabilize.

Another relevant feature of the BMS is ensuring the balancing of the battery cells' charge. Due to slight differences in construction and internal parameters, which increase with charge–discharge cycles and battery lifetime, the energy storage capacity may differ from cell to cell. These differences cause imbalances between the SoC of the cells during normal battery use. Some cells achieve a higher SoC than others, and when a cell with a lower SoC reaches its minimum voltage threshold, the battery stops supplying power even if the remaining cells still have energy to prevent damaging the cell with the lowest SoC [9]. This results in lower battery autonomy, as there is still energy in the battery, but it cannot be used because the cell has reached the minimum voltage limit. On the other hand, during charging, the cells that retained some energy during discharge will charge faster than those that were more discharged, preventing these cells from reaching 100% SoC, thereby creating a cyclical imbalance that exacerbates the issue.

To mitigate this problem, it is necessary to balance the cells to ensure they maintain an equal charge level, thus maximizing battery autonomy [10]. However, the potential of cell balancing has not been well explored, as most BMSs on the market use passive balancing techniques based on dissipating excess energy from the most charged cells through a resistor until they reach the same SoC as the lowest SoC cell [11]. Dissipating energy from the highest SoC cells through resistors is an inefficient and wasteful method of balancing the cells, highlighting the need and opportunity to improve BMSs, particularly regarding balancing circuits [12].

Concerning the research on active balancing circuits for battery cells, the topologies are usually classified according to the main element used in the energy transfer as inductor-based, capacitor-based, and transformer-based topologies [6].

Regarding inductor-based topologies, in [13], the authors propose a balancing circuit where the inductor is the main energy storage element. The circuit, composed of six switches, four capacitors, and three inductors, allows the balance of four cells in a series configuration. An active balancing circuit based on a switched-inductor buck-boost converter with the ability to balance non-adjacent cells is presented in [14]. For a module with three cells in series, the circuit uses two inductors and four switches. In [15], a two-stage active cell-balancing circuit based on a buck-boost converter is proposed. It uses modules consisting of three cells, two inductors, and four MOSFET switches, allowing module-to-module balancing in the first stage and balancing between two cells in a module in the second stage. The authors of [16] propose a new active flying inductor balancing circuit that enables direct energy transfer between any two cells, requiring only one inductor as an intermediate storage element. In [17], the authors propose a new chain-structured cell-balancing circuit using coupled-inductor-based modules achieving high cell-balancing speed.

In terms of capacitor-based topologies, the authors of [18] propose a novel switched-capacitor cell-balancing circuit with a resonant design, allowing all switches employed in the balancing circuit to operate under zero-current switching, thus improving efficiency and performance. The work in [19] proposes an active balancing circuit with two operating modes: any cell to any cell (AC2AC) and direct cell to cell (DC2C). The AC2AC mode is equivalent to a switched capacitor (SC), and the energy is transferred among all imbalanced cells. The DC2C mode is equivalent to a three-resonant-state LC unit, allowing energy transfer from the highest voltage cell to the lowest voltage one.

Transformer-based topologies, also known as isolated topologies, use transformers as the main element in energy transfer. The work in [20] proposes an enhanced switching pattern for an active balancing circuit based on a multi-winding transformer. The circuit, operated in flyback and buck-boost modes, allows the transfer of energy between arbitrary cells. The authors of [21] propose a near-field coupling wireless active cell balancing circuit using planar coupled inductors for high-capacity and high-power battery applications. The work in [22] proposes a novel active cell balancer that supports bidirectional cell balancing operation with very high efficiency, based on an isolated bidirectional dc-dc converter-based active cell balancer. A robust and scalable active battery management system is proposed in [23]. The circuit uses an isolation transformer with two diodes connected to each cell in the battery stack, enabling both accurate cell voltage monitoring and active cell balancing. A cell-to-cell balancing circuit with a bidirectional flyback converter is proposed in [24]. The primary side of the bidirectional flyback converter is connected to the battery pack, and the secondary side is connected to each cell, allowing it to discharge the cell with the highest voltage and charge the cell with the lowest voltage simultaneously.

This paper proposes a new BMS topology for actively balancing battery cells. Unlike other BMSs in the literature, the proposed topology allows the exchange of energy between any cell in the pack as well as with the auxiliary battery of the EV. The main contributions of the proposed BMS topology are:

- Direct energy transfer from the highest SoC to the lowest SoC cell without the involvement of the remaining battery cells.
- Ability to transfer energy between different combinations of cell pairs. For example, at a given moment, n highest SoC cells can transfer energy to m lowest SoC cells ($n, m \in \mathbb{N}$ && $n, m > 0$ && $n + m \leq Total\_of\_cells$).
- Capability to transfer energy between the cells and the auxiliary battery, ensuring it remains charged. It allows, for instance, that only the highest SoC cells contribute to the auxiliary battery, the number of involved cells increasing or decreasing based on demand.
- Option of transferring energy from the service battery to the cells. While this functionality may not be relevant in EVs, it could be useful in other contexts.

All these features significantly contribute to more efficient balancing, opening the possibility of applying different balancing strategies that can be optimized based on the initial SoC or vehicle behavior history, allowing for improved battery autonomy.

This work is organized as follows: Section 1 introduces BMS technology and reviews the literature. Section 2 presents the proposed topology and describes its working principle and functionalities. Section 3 presents a simulation analysis of the proposed topology. Section 4 presents and discusses the experimental results obtained with a small-scale developed prototype implementing the proposed topology. Finally, Section 5 presents the main achievements and conclusions of the work.

2. New BMS Topology

BMS balancing circuits can be classified into two groups: passive balancing and active balancing. Passive balancing circuits are easy and simple to implement; however, they are inefficient in terms of energy conservation. These circuits consist of connecting a resistor in parallel with the highest SoC cells, a topology known as fixed resistor balancing, or switching a resistor in parallel with the highest SoC cells, which allows better control on the dissipated power, a topology known as switched resistor balancing.

Regarding active balancing, which consists of transferring energy from cells with a larger SoC to cells with a smaller SoC, the number of available topologies is very large, with several of them having been discussed in Section 1. Figure 1 clearly demonstrates the energy benefits that can come from active balancing compared to passive balancing. In this illustrative figure, a reduced number of cells are used to highlight the superiority of active balancing over passive balancing. The initial SoC of the cells is: Cell A 20%, Cell B 60%, Cell A 80%, and Cell A 80%. As can be seen, by applying passive balancing, the final SoC of the cells is 20%, whereas by applying active balancing, the final SoC is 60%, which is a significant difference.

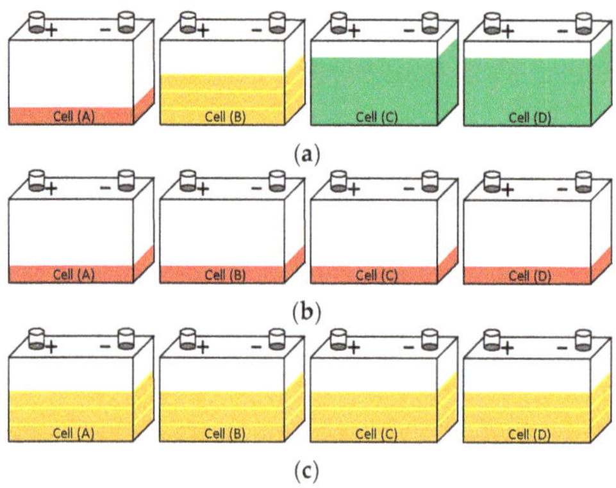

Figure 1. Passive versus active balancing: (**a**) Initial SoC before balancing; (**b**) SoC after passive balancing; (**c**) SoC after active balancing.

Observing Figure 1, it is intuitive to conclude that the strategy for obtaining more efficient balancing consists of transferring 20% SoC from cells C and D to cell A, keeping cell B unchanged. However, most active balancing circuits only allow the exchange of energy between neighboring cells, making it impossible to implement the most efficient balancing strategy. Thus, for the situation presented in Figure 1, balancing would involve transferring energy from cell B to A, from cell C to B, and finally, from cell D to C, repeating the process until all cells have the same SoC. This involves a high set of energy conversions, causing part of the energy that can be used to be wasted during the conversions carried out until balancing is complete. This limitation ends up being a disadvantage of these simpler active balancing topologies.

The proposed topology aims to overcome these limitations by creating a common dc-bus that allows for, at any time, the ability to define which batteries are involved in the balancing process and control the energy flow between them. In electric vehicle applications, the common bus can be connected to the auxiliary battery, greatly improving system functionality. In other applications, the bus can be connected to one of the cells, maintaining the main functionalities. It is important to note that for all cells to be able to exchange energy with the bus, a galvanically isolated bidirectional dc-dc converter is required.

Figure 2 presents a block diagram illustrating the proposed BMS topology and a possible active balancing strategy based on the initial SoC of the cells. In the figure, the arrows represent the energy flow in each point of the circuit, with their thickness proportional to the energy transfer rate (active power). As shown, the cells are connected in series and isolated from the dc-bus and auxiliary battery through bidirectional isolated dc-dc converters. This configuration allows each cell at each instant to inject or absorb energy from the dc-bus independently. The energy in the auxiliary battery is not directly controlled but can be regulated indirectly by operation of each cell's converter. When the total power injected into the dc-bus exceeds the total power absorbed, the excess energy follows to the auxiliary battery. Conversely, if the power injected equals the power absorbed, there is no energy transfer, i.e., the dc-bus voltage equals the auxiliary battery voltage, and the current is zero, resulting in zero active power. A key advantage of the proposed topology is its ability to charge the auxiliary battery or directly supply the loads connected to it. When the current injected by the cells into the dc-bus matches the load current, the auxiliary battery neither charges nor discharges, maintaining a constant SoC. Furthermore, in this case, balancing a high-voltage cell directly with the auxiliary battery requires only a single energy conversion, being twice as efficient as traditional cell-to-cell balancing, which involves two energy conversions (Cell B to dc-bus and dc-bus to Cell A).

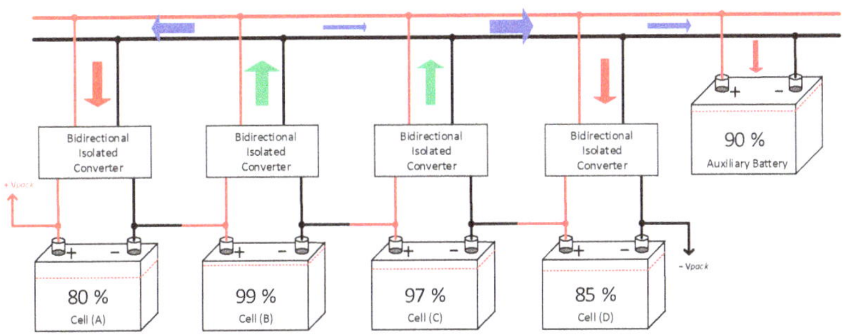

Figure 2. Illustration of the proposed BMS topology architecture with active cell balancing strategy.

2.1. Dc-Dc Bidirectional Isolated Converter

The isolated dc-dc converter in the proposed BMS topology enables energy transfer between the individual cells and the common dc-bus. For active balancing to be effective, the converter must operate with high efficiency, minimizing losses during energy redistribution. Additionally, it should be compact and inexpensive to maintain the overall size and cost of the BMS acceptable.

Among the topologies studied, the bidirectional flyback converter was selected due to its simplicity, suitability for the desired power range, and low component count. These features make it particularly attractive for the BMS under development. Figure 3 shows the schematic of the bidirectional flyback converter used in the active cell balancing circuit. As show, the converter consists of a flyback transformer, an electromagnetic element consisting of two isolated inductors sharing the same magnetic core, two MOSFETs with integrated reverse diodes, and two capacitors. In other words, the low component count enables a compact and cost-effective design. It should be noted that the capacitors are not essential; they are only intended to mitigate current spikes in the cells.

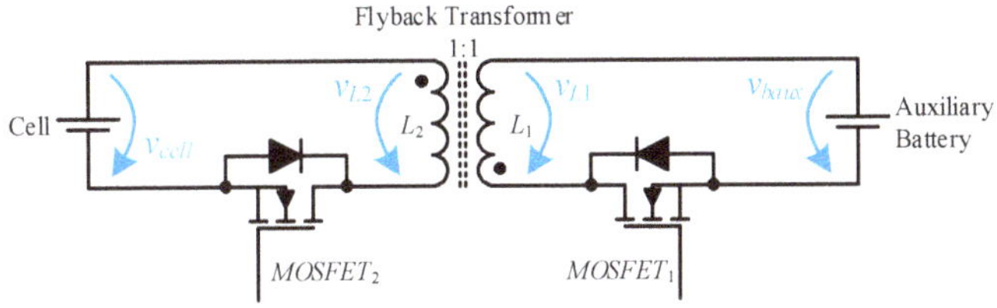

Figure 3. Bidirectional flyback converter used in the proposed BMS for active cell balancing.

The operation of the bidirectional flyback converter is straightforward. To transfer energy from the cell to the auxiliary battery, a pulse width modulation (PWM) signal must be applied to $MOSFET_2$, while $MOSFET_1$ remains off. The power increases with the duty cycle of the PWM signal. For energy transfer in the opposite direction, the process is reversed, $MOSFET_2$ is turned off, and a PWM signal is applied to $MOSFET_1$.

The converter's operation is symmetrical, so a more detailed analysis of its operation will only be carried out in the process of transferring energy from the cell to the auxiliary battery, with $MOSFET_1$ always being turned off and only its internal diode intervening in the process. Thus, assuming all components are ideal and a turn ratio of 1:1 in the flyback transformer, the following can be stated:

- $MOSFET_2$ ON (t_{ON})—When $MOSFET_2$ is turned ON, $v_{L2} = v_{cell}$ and the inductor current increases at a constant ratio, $i_{L2} = \frac{1}{L_2}\int v_{cell}\, dt$. Assuming v_{cell} constant during t_{ON}, then $i_{L2} = \frac{v_{cell} t_{ON}}{L_2}$, and the energy stored in the inductor increases at a constant rate, reaching the final value of $E_{L2} = \frac{(v_{cell} t_{ON})^2}{2L_2}$. During t_{ON}, considering the polarity of the flyback transformer winding, $v_{L1} = -v_{cell}$, causing the $MOSFET_1$ diode to be reverse biased, preventing the circulation of current in L_1.
- $MOSFET_2$ OFF (t_{OFF})—When $MOSFET_2$ is turned OFF, the path for current i_{L2} is interrupted and the L_2 inductor voltage instantaneously goes negative $v_{L2} = -v_{baux}$. A negative voltage on L_2 causes a positive voltage to appear on L_1, $v_{L1} = v_{baux}$, directly biasing the diode of $MOSFET_1$, allowing the energy stored in the flyback transformer to flow to the auxiliary battery. During t_{OFF}, the current in L_1, which instantly reaches the maximum value, decreases at a constant rate as the energy stored in the transformer decreases.

These two states are repeated, ensuring a continuous flow of energy from the cell to the auxiliary battery. Two scenarios can occur depending on the duty cycle and the frequency of the PWM signal.

Firstly, if all the energy stored in the flyback transformer is discharged into the auxiliary battery and, when $MOSFET_2$ has turned ON again, the current starts from zero, then the converter will be operating in discontinuous mode. In this mode, the converter efficiency will be better, and the stress on the semiconductors will be lower.

Secondly, if energy remains stored in the transformer when $MOSFET_2$ is turned on, the current will increase from an initial value greater than zero, and the converter will operate in continuous mode. In this mode, the power density is higher, but the stress on the semiconductors increases.

This work explores the converter in discontinuous mode, though both modes are possible within the proposed concept. Further details on the converter's operation will be presented in Sections 3 and 4, based on simulation and experimental results.

2.2. Distributed Architecture

This work aimed to develop a modular BMS topology that is scalable for a large number of cells. Therefore, a distributed architecture based on a master–slave hierarchy was chosen. The design features a master module connected to the auxiliary battery, which performs most of the typical BMS functions, including monitoring the battery pack current and providing features, such as overcurrent protection, Coulomb counting, SoC estimation, and CAN communication with the other EV subsystems. The master is also responsible for monitoring the voltage and temperature of the auxiliary battery, as well as receiving information about the voltage and temperature of each cell through a digital communication bus. For this work, the use of a slave for each cell was implemented, utilizing a RS485 bus communication. The RS485 was chosen to avoid the need for more expensive microcontrollers with built-in CAN peripherals.

Regarding the BMS architecture, many other options could have been considered, such as using a slave module for a set of 4 or 5 cells, using other communication technologies, etc. However, this is not the main focus of this work; the objective was to obtain a flexible and functional BMS topology. Figure 4 shows a simplified schematic of the distributed BMS topology where the master and three slaves are represented.

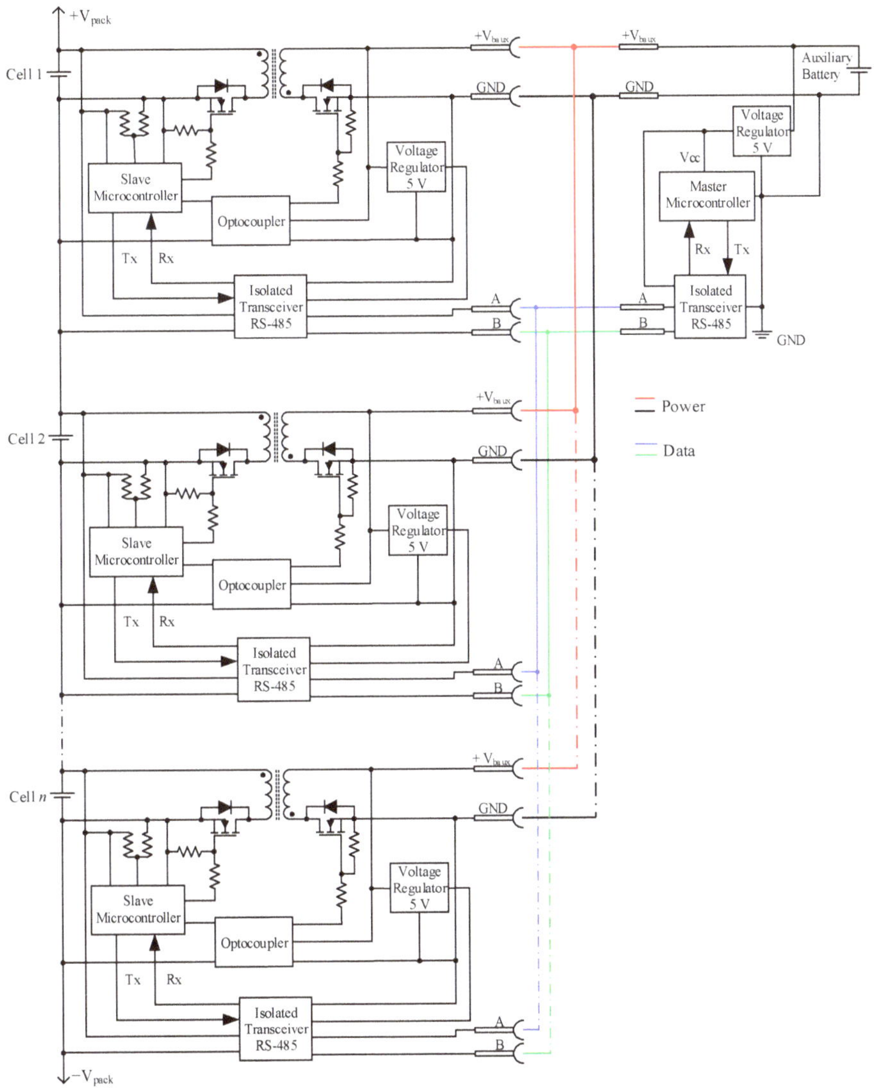

Figure 4. Simplified schematic of the BMS distributed topology.

The master microcontroller powered by the auxiliary battery is responsible for the entire operation of the system. On the other hand, the slave microcontrollers, powered by the cells, communicate with the master through a galvanically isolated RS485 bus, providing the master with information about the status of the monitored cell and receiving orders to send or receive energy from the balancing circuit.

The PWM signals applied to the converter MOSFETs are generated locally by the slave microcontrollers. During the project, several precautions were taken to ensure the proper operation of the microcontrollers throughout the entire voltage range of the cells. More details on the selected components are presented in Section 4.

The topology proposed in this work allows energy transfer between any cell to any cell through a dc-bus connected to the auxiliary battery of the EV, enabling the energy exchange between the traction battery (high power, high voltage) and the auxiliary battery (low power, low voltage). This flexibility in energy exchange supports a wide range of balancing strategies. The greater the number of cells, the greater the number of balancing combinations. Appendix A presents and describes an example code synopsis for a possible balancing strategy.

3. Simulation of Dc-Dc Bidirectional Isolated Converter

This section presents the results obtained through computer simulations, developed with PSIM 9.1 software, related to the dc-dc bidirectional isolated converter. Nowadays, simulation is an important step in the sizing of power electronics converters and in improving control algorithms. If the converter parameters are modeled accurately and the simulations are conducted properly, the results obtained serve as a good indicator of what the converter's behavior will be experimentally.

For this work it was defined that the cells to be used would be of lithium iron phosphate ($LiFePO_4$) technology with a 3.2 V nominal voltage, and the auxiliary battery would be of lead-acid technology with a nominal voltage of 12.8 V. It was also defined that the balancing current would range between 0.1 and 1 A.

Based on these parameters, a search was carried out for flyback transformers best suited for this application. A transformer with three windings, one primary and two secondaries, was selected: the WE-FLYLT 750313973 from Würth Elektronik (Würth Elektronik eiSos GmbH & Co. KG, Waldenburg, Germany). Figure 5 shows the schematic of the flyback transformer windings arrangement, pinout, and typical operation characteristics. The main electrical specifications of the flyback transformer are presented in Table 1. In this work, the winding N_1 is connected to the high-side voltage (auxiliary battery), and the windings N_2 and N_3 are connected in parallel to the low-side voltage (cell), effectively forming a single winding.

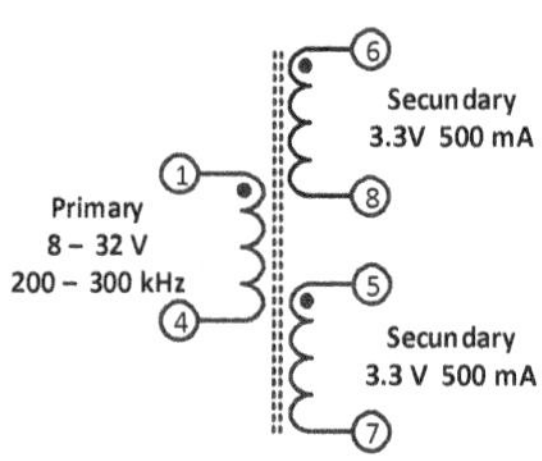

Figure 5. Flyback transformer windings arrangement and pinout.

Table 1. Electrical properties of the flyback transformer WE-FLYLT 750313973.

Parameter	Test Conditions	Value
Inductance, L (1–4)	100 kHz, 100 mV	40 µH, ±10%
Turns ration, n	N1:N2:N3	4:1:1
Saturation current (1–4)	$\|\Delta L/L\| < 20\%$	2 A, typ.
Dc resistance, R_{DC1} (1–4)	20 °C	0.08 Ω, max.
Dc resistance, R_{DC2} (6–8)	20 °C	0.04 Ω, max.
Dc resistance, R_{DC3} (5–7)	20 °C	0.04 Ω, max.
Leakage inductance, Ls (1–4)	100 kHz, 100 mV	1 µH, max.
Insulation test voltage, V_T	From N1 to N2, N3	1500 V (ac)

To simulate the flyback transformer, the "coupled inductor" component was used. Considering L_{11} and L_{22} as self-inductances of the primary and secondary windings, respectively, and L_{12} and L_{21} as the mutual inductances, the voltages and currents in the primary (v_1, i_1) and secondary (v_2, i_2) windings are related as follows:

$$\begin{bmatrix} v_1 \\ v_2 \end{bmatrix} = \begin{bmatrix} L_{11} & L_{12} \\ L_{21} & L_{22} \end{bmatrix} \frac{d}{d_t} \begin{bmatrix} i_1 \\ i_2 \end{bmatrix}. \quad (1)$$

Considering mutual inductances of equal value, i.e., $L_{12} = L_{21}$, the transformer coupling factor, k, can be calculated by:

$$k = \frac{L_{12}}{\sqrt{L_{11}L_{22}}} \Longleftrightarrow L_{12} = k\sqrt{L_{11}L_{22}}. \quad (2)$$

The coupling factor gets a value between 0 and 1. The higher the coupling factor, the better, and in the limit, 1 represents ideal coupling.

The self-inductance values are proportional to the square of the number of turns. Therefore, if N_1 is the primary winding turns number and N_2 the secondary winding turns number, then the relation between the self-inductances can be obtained by:

$$\frac{L_{11}}{L_{22}} = \frac{N_1^2}{N_2^2} \quad (3)$$

Knowing that $L_{11} = 40$ µH (from Table 1), L_{22} will be 2.5 µH, given that the transformer ratio is 4:1. The relation between the primary sell inductance (L_{11}) and the primary leakage inductance (L_{11}^{σ}) is obtained by (4), where k is the windings coupling factor.

$$L_{11}^{\sigma} = L_{11}(1 - k) \quad (4)$$

Using $L_{11}^{\sigma} = 1$ µH, (from Table 1), the coupling factor value of $k = 0.975$ was obtained. Thus, the value of mutual inductance can be calculated, $L_{12} = L_{21} = 9.75$ µH. The transformer parameters for the simulation are summarized in Table 2.

Table 2. Flyback transformer simulation parameters.

Parameter	Value
Primary winding inductance, L_{11}	40 µH
Secondary winding inductance, L_{22}	2.5 µH
Turns ratio, n	4:1
Mutual inductance, $L_{12} = L_{21}$	9.75 µH
Primary winding dc resistance, R_{DC1}	0.08 Ω
Secondary winding dc resistance, R_{DC2}	0.02 Ω

Figure 6 presents the schematic of the bidirectional flyback converter used in the simulation. Unlike the circuit presented in Figure 3, this circuit accounts for the components' parasitic elements to achieve more realistic results. Table 3 presents the values of the components used in the simulation model.

Analyzing the operation in the energy transfer mode from the cell to the auxiliary battery, $MOSFET_1$ is always OFF, and two different states arise depending on the state of $MOSFET_2$.

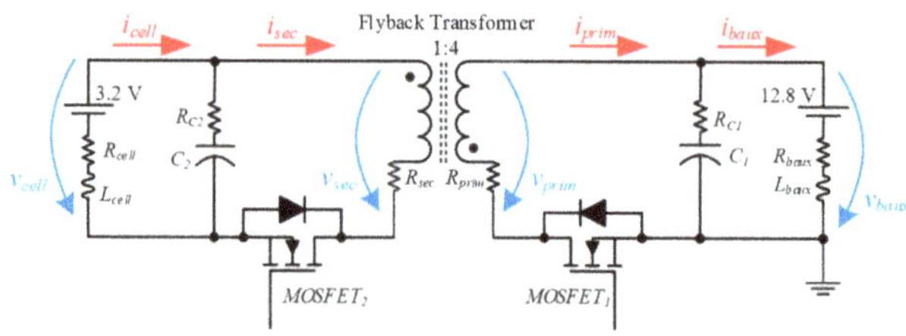

Figure 6. Schematic of the bidirectional flyback converter used in the simulation.

Table 3. Simulation parameters.

Parameter	Value
Primary capacitor, C_1	100 µF
Equivalent series resistor (ESR) of C_1, R_{C1}	5 mΩ
Secondary capacitor, C_2	100 µF
Equivalent series resistor (ESR) of C_2, R_{C2}	5 mΩ
Cell internal resistance, R_{cell}	2 mΩ
Cell internal inductance, L_{cell}	10 nH
Auxiliary battery internal resistance, R_{baux}	8 mΩ
Auxiliary battery internal inductance, L_{baux}	10 ηH
Switching frequency, f_s	250 kHz

Figure 7 represents the two states mentioned, where Figure 7a shows the components involved in the energy transfer from the cell to the flyback transformer during t_{ON}, and Figure 7b shows the components involved in the energy transfer from the flyback transformer to the auxiliary battery during t_{OFF}. Greyed-out components are not used in these states.

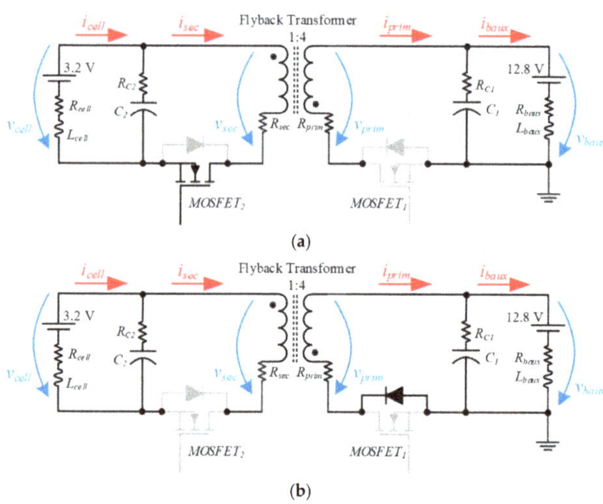

Figure 7. Bidirectional flyback converter transferring energy from the cell to the auxiliary battery: (**a**) During t_{ON}, when $MOSFET_2$ is ON; (**b**) During t_{OFF}, when $MOSFET_2$ is OFF.

Figure 8 presents the simulation results of the bidirectional flyback converter transferring energy from the cell to the auxiliary battery. Figure 8a shows the pulse width modulation (PWM) signal of $MOSFET_2$. Figure 8b shows the secondary winding voltage (v_{sec}) when $MOSFET_2$ is ON, v_{sec} is equal to v_{cell}, and v_{prim} is negative with four times more amplitude. When the $MOSFET_2$ is turned OFF, v_{sec} becomes negative with a quarter of the auxiliary battery voltage amplitude, and v_{prim} is positive and equal to the auxiliary battery voltage. The winding voltages become zero when all the energy stored in the transformer is transferred to the auxiliary battery. This demonstrates that the converter operates in discontinuous mode.

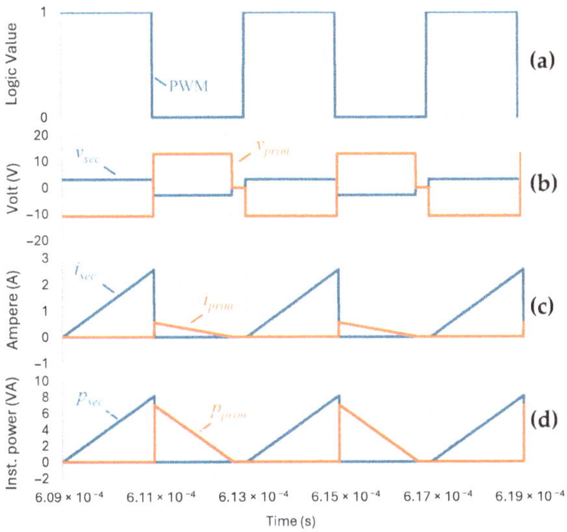

Figure 8. Simulation results of the bidirectional flyback converter transferring energy from the cell to the auxiliary battery: (**a**) PWM signal of $MOSFET_2$; (**b**) Voltages in the transformer secondary and primary windings; (**c**) Currents in the transformer secondary and primary windings; (**d**) Instantaneous power in the transformer secondary and primary windings.

Figure 8c shows the secondary and primary currents. When $MOSFET_2$ is turned ON, i_{sec} increases at a constant rate, starting from zero. When the $MOSFET_2$ is turned OFF, i_{sec} ceases and i_{prim} instantaneously reaches its maximum value before decreasing at a constant rate until zero. The maximum value of i_{sec} is four times greater than the maximum value of i_{prim}, due to the transformation ratio.

Figure 8d shows the instantaneous power in the transformer's secondary and primary windings. The area of the trapezoid, p_{sec}, corresponds to the energy transferred from the cell to the flyback transformer. The area of the trapezoid, p_{prim}, corresponds to the energy transferred from the flyback transformer to the auxiliary battery and is positive because of the assumed direction of the auxiliary battery current (see Figure 6).

These results demonstrate the correct operation of the bidirectional flyback transformer transferring energy from the cell to the auxiliary battery. As the converter operates symmetrically, the results of transferring energy from the auxiliary battery to the cell are redundant and are not presented.

4. Experimental Validation of the Dc-Dc Bidirectional Isolated Converter

This section presents the results obtained with the BMS prototype developed using the proposed topology. The experimental setup used to acquire the results consists of three 3.2 V, 100 Ah, $LiFePO_4$ cells connected in series (one per slave module). A 12 V, 45 Ah lead-acid battery was used to emulate the EV's auxiliary battery. A four-channel Tektronix TPS 2024B oscilloscope (Tektronix, Beaverton, OR, USA) and a Tektronix A6302 current clamp were used to capture the results.

Figure 9 presents the BMS prototype developed for experimental tests. Figure 9a shows the master PCB, Figure 9b shows the slave PCB, Figure 9c shows the prototype assembled with three series of 3.2 V, 100 Ah $LiFePO_4$ cells, and a 12 V lead-acid battery, ready for tests, and Figure 9d shows a detail of current and voltage probes used during experimental tests.

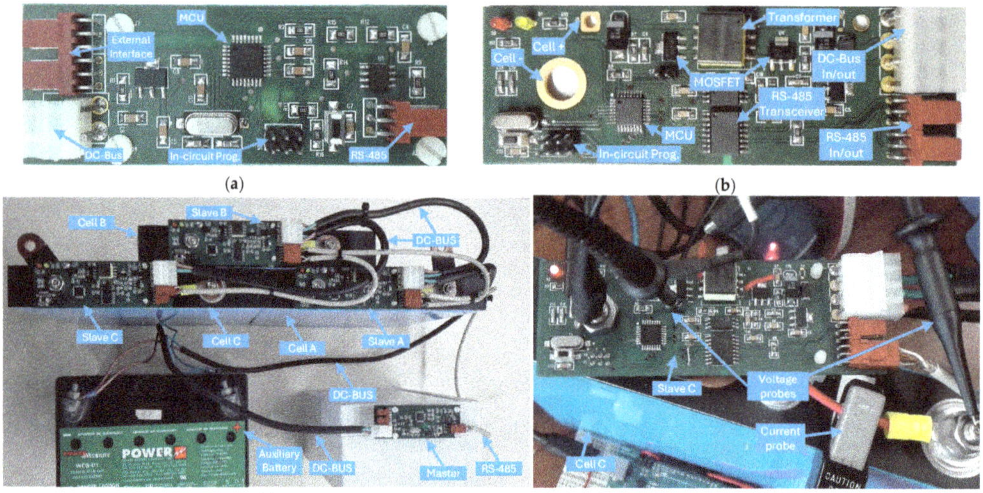

(a) (b) (c) (d)

Figure 9. Developed laboratory prototype of the BMS: (**a**) Master module; (**b**) One of the three slave modules; (**c**) BMS assembled with three cells and auxiliary battery; (**d**) Detail of current and voltage probes used during experimental results acquisition.

4.1. Energy Transfer from the Cell to the Auxiliary Battery

During energy transfer from the cell to the auxiliary battery, the $MOSFET_1$ is kept OFF, and a PWM signal is applied to $MOSFET_2$. The energy transfer occurs in two steps: during t_{ON}, the energy is transferred from the cell to the flyback transformer; during t_{OFF}, the energy is transferred from the flyback transformer to the auxiliary battery.

Figure 10 presents experimental results during the energy transfer from the cell to the auxiliary battery. The results in Figure 10a refer to the secondary side of the flyback transformer and show the energy transfer from the cell to the flyback transformer. During t_{ON}, the secondary winding voltage (v_{sec}) is equal to the cell voltage (v_{cell}), the secondary current (i_{sec}) increases almost linearly, and the instantaneous power (p_{sec}) has similar behavior to i_{sec}. During t_{OFF}, the secondary winding voltage (v_{sec}) is equal to $-$¼ of the auxiliary battery voltage (v_{baux}), the secondary current (i_{sec}) is zero, and, as a result, the instantaneous power (p_{sec}) is also zero. As expected, the cell voltage (v_{cell}) does not undergo any changes during the two-time intervals.

The results in Figure 10b refer to the primary side of the flyback transformer and show the energy transfer from the flyback transformer to the auxiliary battery. During t_{ON}, the primary winding voltage (v_{prim}) is equal to $-4\ v_{cell}$, the primary current (i_{prim}) is zero, and the instantaneous power (p_{prim}) is also zero. During t_{OFF}, the primary winding voltage (v_{prim}) is equal to the auxiliary battery voltage (v_{baux}), the primary current (i_{prim}) instantaneously reaches its maximum value and decreases almost linearly, and the instantaneous power (p_{prim}) follows a similar behavior to i_{prim}. As expected, the auxiliary battery voltage (v_{baux}) does not undergo any changes during the two-time intervals.

The results obtained when transferring energy from the cell to the auxiliary battery correspond to what was expected. The waveforms obtained from both the primary and secondary sides of the flyback transformer are very similar to the simulation results, with only minor differences due to the non-ideal nature of the real electronic components and limitations of the measuring instruments.

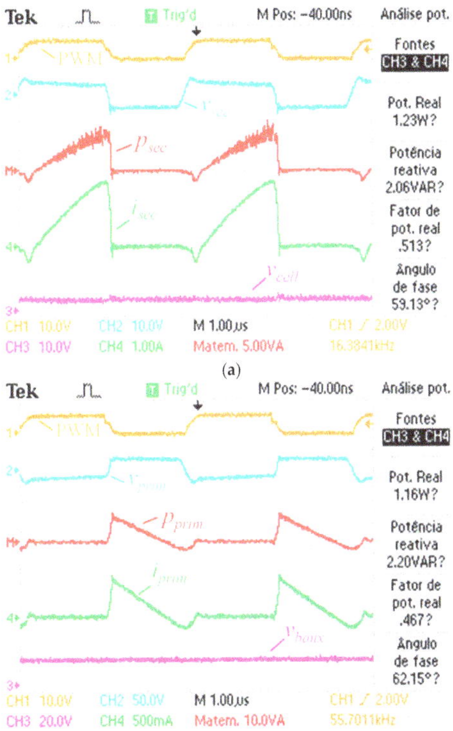

Figure 10. Experimental results of the BMS prototype transferring energy from the cell to the auxiliary battery: (**a**) PWM signal applied to $MOSFET_2$, voltage in the secondary winding of flyback transformer, current in the secondary winding of flyback transformer, cell voltage, and instantaneous power in the secondary winding of flyback transformer; (**b**) PWM signal applied to $MOSFET_2$, voltage in the primary winding of flyback transformer, current in the primary winding of flyback transformer, auxiliary battery voltage, and instantaneous power in the secondary winding of flyback transformer.

4.2. Energy Transfer from the Auxiliary Battery to the Cell

During the energy transfer from the auxiliary battery to the cell, $MOSFET_2$ is kept OFF, and a PWM signal is applied to $MOSFET_1$. The energy transfer occurs in two steps: during t_{ON}, the energy is transferred from the auxiliary battery to the flyback transformer; during t_{OFF}, the energy is transferred from the flyback transformer to the cell.

Figure 11 presents the experimental results during the energy transfer from the auxiliary battery to the cell. The results in Figure 11a refer to the primary side of the flyback transformer and show the energy transfer from the auxiliary battery to the flyback transformer. During t_{ON} the primary winding voltage (v_{prim}) is equal to the auxiliary battery voltage (v_{baux}), the primary current (i_{prim}) decreases almost linearly, and the instantaneous power (p_{prim}) has similar behavior to i_{prim}. The reason for obtaining negative values for i_{prim} and p_{prim} is the direction assumed as positive for i_{prim} (see Figure 6).

The results in Figure 11b refer to the secondary side of the flyback transformer and show the energy transfer from the flyback transformer to the auxiliary battery. During t_{ON} the secondary winding voltage (v_{sec}) is equal to $-¼\ v_{baux}$, the secondary current (i_{sec}) is zero, and consequently, the instantaneous power (p_{sec}) is also zero. During t_{OFF}, the secondary winding voltage (v_{sec}) is equal to v_{cell}, the secondary current (i_{sec}) instantaneously decreases until the minimum value and then increases almost linearly until zero, and the instantaneous power (p_{sec}) has similar behavior to i_{sec}.

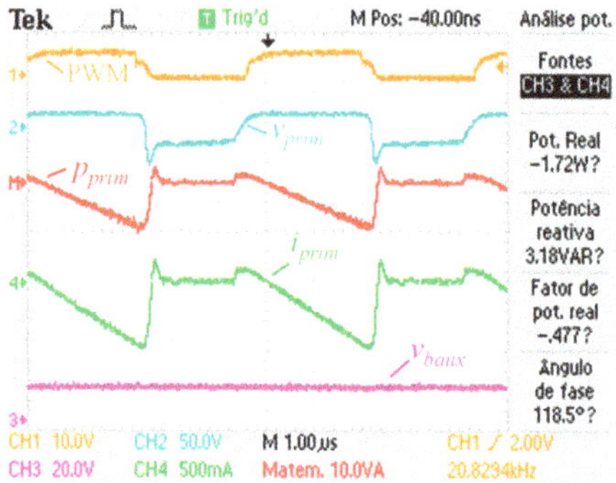

(a)

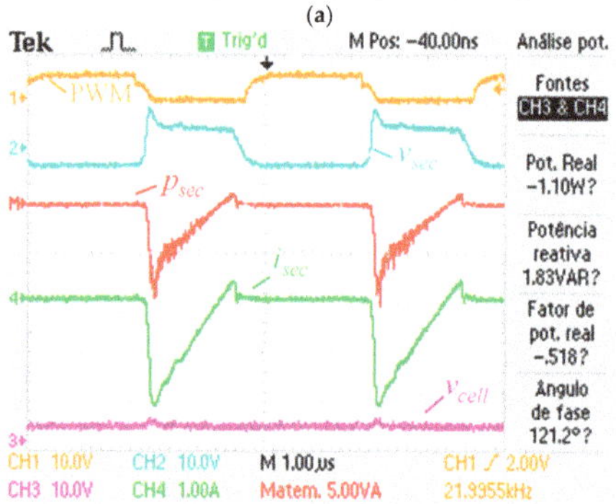

(b)

Figure 11. Experimental results of the BMS prototype transferring energy from the auxiliary battery to the cell: (**a**) PWM signal applied to $MOSFET_1$, voltage in the primary winding of flyback transformer, current in the primary winding of flyback transformer, auxiliary battery voltage, and instantaneous power in the primary winding of flyback transformer; (**b**) PWM signal applied to $MOSFET_1$, voltage in the secondary primary of flyback transformer, current in auxiliary battery, auxiliary battery voltage, and instantaneous power in the auxiliary battery.

As before, the results obtained when transferring energy from the auxiliary battery to the cell align with the expectations. The results in this mode are practically symmetrical to the signals obtained in the previous mode. Once again, the waveforms presented highlight the non-ideal nature of the real electronic components and limitations of the measuring instruments.

Figure 12 presents experimental results during the energy transfer from the cell to the auxiliary battery to assess power conversion efficiency.

The results in Figure 12a refer to the secondary side of the flyback transformer and show the energy transfer from the cell to the flyback transformer. During t_{ON}, the secondary winding voltage (v_{sec}) is equal to the cell voltage (v_{cell}), the cell current (i_{cell}) increases, and the instantaneous power (p_{sec}) has similar behavior to i_{cell}. During t_{OFF}, the secondary winding voltage (v_{sec}) is equal to −¼ of the auxiliary battery voltage (v_{baux}), the secondary current (i_{sec}) is zero, but the cell continues to charge the capacitor C_2 (Figure 7) and, as a result, the cell current decreases as the capacitor voltage increases. As expected, the cell voltage (v_{cell}) does not undergo any changes during the two-time intervals.

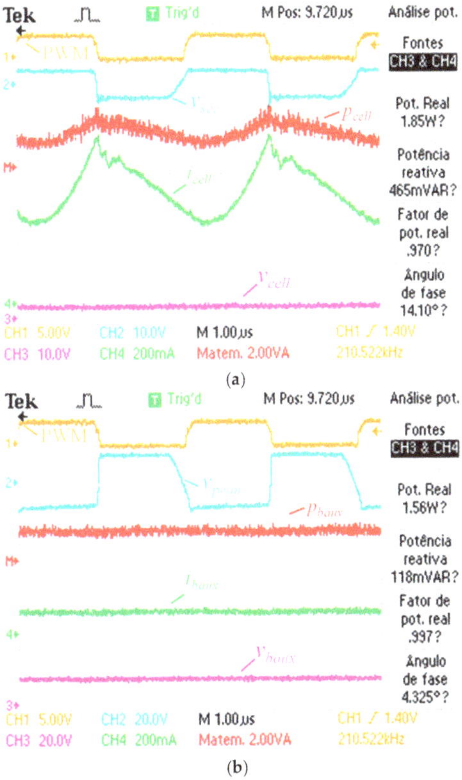

Figure 12. Experimental results of the BMS prototype transferring energy from the cell to the auxiliary battery: (**a**) PWM signal applied to $MOSFET_2$, voltage in the primary winding of flyback transformer, current in the cell, cell voltage, and instantaneous in the cell; (**b**) PWM signal applied to $MOSFET_2$, voltage in the secondary winding of flyback transformer, current in the auxiliary battery, auxiliary battery voltage, and instantaneous power in the auxiliary battery.

The results in Figure 12b refer to the primary side of the flyback transformer and show the energy transfer from the flyback transformer to the auxiliary battery. During t_{ON}, the primary winding voltage (v_{prim}) is equal to $-4\ v_{cell}$. As a consequence of capacitor C_2 (Figure 7) and the high impedance of the dc-bus wiring acting as a low-pass filter, the auxiliary battery current (i_{baux}) is almost constant, and the instantaneous power in the auxiliary battery (p_{baux}) has similar behavior. As expected, the auxiliary battery voltage (v_{baux}) does not undergo any changes during the two-time intervals.

Considering the active power measured by the oscilloscope, the efficiency of the power converter can be calculated as 0.843 (1.56/1.85), almost 84%.

The experimental results obtained confirm the flexibility and reliability of the bidirectional flyback converter in transferring energy both from the cell to the auxiliary battery and in the opposite direction, proving to be a valid choice for the distributed BMS architecture with active balancing. The measured power conversion efficiency was quantified as 84%. The authors believe that with an increase in the dc-bus wiring section and some improvements to the circuit such as in the layout, the adoption of an external diode, Schottky type, to be placed in parallel with the internal diodes of the $MOSFET_1$ and $MOSFET_2$, it will be possible to obtain an efficiency greater than 90%. Another aspect to be improved in a new version of the BMS prototype relates to safety and fault protection. The adoption of fuses should be considered in the event of insulation failure in the flyback transformer to isolate the failure.

Regarding communications, the use of an optical fiber ring topology, or even a wireless topology, could be considered but at an additional cost. Finally, a technique to enhance the resistance of the circuits against external factors is to immerse them in epoxy resin.

5. Conclusions

This paper proposes a new BMS topology with distributed architecture, active cell balancing, and the capability to exchange energy between any cells in the traction pack and with the EV auxiliary battery. The proposed topology is based on a master/slave architecture where the master module monitors the auxiliary battery, and each of the slave modules monitors one of the pack cells. All the models are connected to a communication bus based on an RS-485 physical layer.

A bidirectional flyback converter was used to transfer energy between each of the pack cells, with a common dc-bus connected to the auxiliary battery, facilitating the exchange of energy between any cell in the traction battery pack and with the auxiliary battery, allowing at each time, the selection of the cells involved in the balancing process, and enabling charging of the auxiliary battery if necessary.

The flexibility of the proposed BMS permits different balancing strategies, improving the balancing efficiency. The active cell balancing circuit was analyzed through computer simulations and tested experimentally in a developed small-scale laboratory prototype. The simulation and experimental results validated the ability of the system to transfer energy between the cells and the auxiliary battery, proving to be a valid architecture for the distributed BMS with active cell balancing. The power conversion efficiency was measured as 84%. Despite the flexibility in energy transfer, the proposed topology presents some limitations and potential disadvantages, such as higher volume, height, and cost, compared to other BMS topologies, namely those with passive balancing.

Author Contributions: Conceptualization, J.G.O.P. and M.F.S.; methodology, J.G.O.P. and M.F.S.; software, M.F.S.; validation, M.F.S., J.G.O.P. and L.A.M.B.; formal analysis, J.G.O.P., M.F.S., L.A.M.B. and J.A.A.; investigation, J.G.O.P. and M.F.S.; resources, J.G.O.P.; data curation, J.G.O.P., M.F.S., L.A.M.B. and J.A.A.; writing—original draft preparation, J.G.O.P.; writing—review and editing, J.G.O.P., M.F.S., L.A.M.B. and J.A.A.; visualization, M.F.S., L.A.M.B. and J.A.A.; supervision, J.G.O.P. All authors have read and agreed to the published version of the manuscript.

Funding: This work has been supported by FCT—Fundação para a Ciência e Tecnologia within the R&D Unit Project Scope UID/00319/Centro ALGORITMI (ALGORITMI/UM).

Data Availability Statement: The original contributions presented in the study are included in the article, further inquiries can be directed to the corresponding authors.

Conflicts of Interest: The authors declare no conflicts of interest.

Abbreviations

The following abbreviations are used in this manuscript:

BMS	battery management system
CAN	controller area network
dc	direct current
DoD	depth of discharge
EV	electric vehicle
MOSFET	metal oxide semiconductor field effect transistor
PWM	pulse width modulation
SoC	state of charge
SoH	state of health

Appendix A

Figure A1 presents an example code synopsis for a balancing strategy applied to a pack of 100 cells, 3.2 V/100 Ah (320 Wh), without involving the auxiliary battery (at the end of balancing, the final energy in the auxiliary battery will be equal to the energy existing at the beginning of balancing) based on OCVM. To implement this strategy, the BMS measures the initial cell voltages to estimate their initial SoC (*cell_SoC[i]*) and the initial total SoC (*total_SoC*). Dividing the initial total SoC by the number of cells gives their average initial SoC (*avg_initial_SoC*). It then calculates the total energy to be transferred during cell balancing (*total_balancing_energy*). Based on this amount of energy, the algorithm estimates the energy losses in the power converter circuits (*losses*), considering two power conversions: from the cell B to the dc-bus and from the dc-bus to cell A. The SoC after balancing (*final_SoC*) will be equal to the initial SoC minus the losses during balancing. Based on the final SoC, it is possible to estimate the cells' balancing end voltage (*target_cell_v*). Then, the system enters a balancing loop, ending when all the cells reach the target voltage. A hysteresis of 1% is applied to prevent the cells from entering consecutive charging/discharging states and a hysteresis of 2% is applied to assume the cell as balanced. This value considers the cell voltage rebound and assumes that the cells are balanced within a tolerance range of 2% related to the target voltage.

```
#define N 100;//number of cells (from 0 to N-1)
#define E_cell 320;//cell energy (Wh)corresponding to 100% SoC
#define Efficiency 0.84;//efficiency of the power converter (90 %)
int start_balancing(void){
    total_SoC=0;
    total_balancing_energy=0;
    balanced_cells=0;
    for (i=0;i<N;i++){
        cell_v=read_cell_voltage(i);//read cell voltage
        cell_SoC[i]=estimate_cell_SoC(cell_v);//estimate cell SoC
        total_SoC+=cell_SoC[i];//calculate total cells SoC
    }
    avg_initial_SoC=total_SoC/N;//calculate cells initial average SoC
    for (i=0;i<N;i++){
        SoC_diff=cell_SoC[i]-avg_initial_SoC;//calculate difference between cell SoC and initial cells average SoC
        total_balancing_energy+=abs(SoC_diff)*E_cell;//total energy (Wh) to be transferred during the balancing
    }
    losses=total_balancing_energy*(1-Efficiency)*2; //energy losses during balancing considering two energy conversions
    final_SoC=total_SoC-(losses/E_cell);//calculate final SoC after balancing
    avg_final_SoC=final_SoC/N;//calculate final average cell SoC after balancing (target SoC)
    target_cell_v=estimate_cell_voltage(avg_final_SoC);//estimate the voltage corresponding to final cell SoC (target voltage)
    while(balanced_cells<N){
        for (i=0;i<N;i++){
            cell_v=read_cell_voltage(i);//read cell voltage
            if (cell_v>target_cell_v*1.01)//considering a 1% margin of hysteresis
            battery_command[i]=1;//discharging state
            else if(cell_v < target_cell_v*0.99)//considering a 1% margin of hysteresis
            battery_command[i]=2; //charging state
            else if ((cell_v>target_cell_v*0.98)&&(cell_v<target_cell_v*1.02)){//1% additional for cell voltage rebound
                battery_comand[i]=0;//off state
                balanced_cells++;//increment number of balanced cells
            }
        }
    }
    return 0; // end of balancing
}
```

Figure A1. Example code synopsis of a balancing strategy for the pack cells without involving the auxiliary battery.

References

1. Tintelecan, A.; Dobra, A.C.; Marţiş, C. LCA Indicators in Electric Vehicles Environmental Impact Assessment. In Proceedings of the 2019 Electric Vehicles International Conference (EV), Bucharest, Romania, 3–4 October 2019; pp. 1–5.
2. Murali, N.; Mini, V.P.; Ushakumari, S. Electric Vehicle Market Analysis and Trends. In Proceedings of the 2022 IEEE 19th India Council International Conference (INDICON), Kochi, India, 24–26 November 2022; pp. 1–6.
3. Roboam, X. A Review of Powertrain Electrification for Greener Aircraft. *Energies* **2023**, *16*, 6831. [CrossRef]
4. Kumar, R.R.; Bharatiraja, C.; Udhayakumar, K.; Devakirubakaran, S.; Sekar, K.S.; Mihet-Popa, L. Advances in Batteries, Battery Modeling, Battery Management System, Battery Thermal Management, SOC, SOH, and Charge/Discharge Characteristics in EV Applications. *IEEE Access* **2023**, *11*, 105761–105809. [CrossRef]
5. Vijaychandra, J.; Knypiński, Ł. A Comprehensive Review on Challenges and Possible Solutions of Battery Management Systems in Electric Vehicles. In Proceedings of the 2024 Progress in Applied Electrical Engineering (PAEE), Koscielisko, Poland, 24–28 June 2024; pp. 1–6.
6. Miranda, J.P.D.; Barros, L.A.M.; Pinto, J.G. A Review on Power Electronic Converters for Modular BMS with Active Balancing. *Energies* **2023**, *16*, 3255. [CrossRef]
7. Li, Z.; Huang, J.; Liaw, B.Y.; Zhang, J. On state-of-charge determination for lithium-ion batteries. *J. Power Sources* **2017**, *348*, 281–301. [CrossRef]
8. Meng, J.; Ricco, M.; Luo, G.; Swierczynski, M.; Stroe, D.-I.; Stroe, A.-I.; Teodorescu, R. An Overview and Comparison of Online Implementable SOC Estimation Methods for Lithium-Ion Battery. *IEEE Trans. Ind. Appl.* **2018**, *54*, 1583–1591. [CrossRef]
9. Chauhan, S.R.; Kumar, K.; Nadarajan, S.; Vaiyapuri, V.; Halick, M.; Sathik, M. Effect of Unbalanced Cells in Lithium-ion Battery Pack Performance and SOC Estimation. In Proceedings of the 2024 12th International Conference on Internet of Everything, Microwave, Embedded, Communication and Networks (IEMECON), Jaipur, India, 24–26 October 2024; pp. 1–6.
10. Pavan, M.V.S.; Ravindranadh, V.; Shareef, I.; Sathish, D.V.S.; Sankar, J.P.S. Optimizing Battery Performance -Active and Passive Cell Balancing. In Proceedings of the 2024 Third International Conference on Intelligent Techniques in Control, Optimization and Signal Processing (INCOS), Krishnankoil, India, 14–16 March 2024; pp. 1–6.
11. Apipatsakul, S.; Fuengwarodsakul, N.; Masomtob, M. Design Guidelines of Passive Balancing Circuit for Li-Ion Battery for Bleeding Current Adjustment Using PWM Technique. In Proceedings of the 2021 Research, Invention, and Innovation Congress: Innovation Electricals and Electronics (RI2C), Bangkok, Thailand, 1–3 September 2021; pp. 273–277.
12. Lim, W.C.; Terence Teo, B.C.; Lim, X.Y.; Siek, L.; Tan, E.L. Adaptive Active Balancing in Battery Management Systems with Comprehensive System Modeling. In Proceedings of the 2025 International Conference on Electronics, Information, and Communication (ICEIC), Osaka, Japan, 19–22 January 2025; pp. 1–5.
13. Panchal, A.; Bhatt, K.; Gitaye, S.; Bhand, M.; Sheikh, A. Design and Simulation of an Inductor based Active Cell Balancing Circuit for Lithium-ion Batteries. In Proceedings of the 2022 4th Global Power, Energy and Communication Conference (GPECOM), Nevsehir, Turkey, 14–17 June 2022; pp. 89–94.
14. Manjunath, K.; Kalpana, R. Active Cell Balancing Circuit using Switched inductor Buck-Boost Converter for Li-ion Battery Strings with Maximum Efficiency Operation. In Proceedings of the 2023 IEEE International Conference on Power Electronics, Smart Grid, and Renewable Energy (PESGRE), Trivandrum, India, 17–20 December 2023; pp. 1–6.
15. Manjunath, K.; Kalpana, R. A Modularized Two-Stage Active Cell Balancing Circuit for Series Connected Li-Ion Battery Packs. In Proceedings of the 2022 IEEE International Conference on Power Electronics, Drives and Energy Systems (PEDES), Jaipur, India, 14–17 December 2022; pp. 1–6.
16. Kipke, V.; Dost, P.; Sourkounis, C.; Widera, B. Novel Direct Active Cell-to-Cell Balancing Approach for Energy Storage Systems based on a Flying Inductor Circuit Topology. In Proceedings of the IECON 2024—50th Annual Conference of the IEEE Industrial Electronics Society, Chicago, IL, USA, 3–6 November 2024; pp. 1–8.
17. Lee, E.-S.; Lee, S.-W.; Kim, C.-S.; Han, J.-K. A New Chain-Structured Cell-Balancing Circuit with a Coupled-Inductor Based Modules. In Proceedings of the 2024 3rd International Conference on Power Systems and Electrical Technology (PSET), Tokyo, Japan, 5–8 August 2024; pp. 429–434.
18. Ye, Y.; Cheng, K.W.E. Analysis and Design of Zero-Current Switching Switched-Capacitor Cell Balancing Circuit for Series-Connected Battery/Supercapacitor. *IEEE Trans. Veh. Technol.* **2018**, *67*, 948–955. [CrossRef]
19. Zhou, G.; Zhang, X.; Gao, K.; Tian, Q.; Xu, S. Two-Mode Active Balancing Circuit Based on Switched-Capacitor and Three-Resonant-State LC Units for Series-Connected Cell Strings. *IEEE Trans. Ind. Electron.* **2022**, *69*, 4845–4858. [CrossRef]
20. Lee, S.; Kim, M.; Baek, J.W.; Kang, D.-W.; Jung, J. Enhanced Switching Pattern to Improve Cell Balancing Performance in Active Cell Balancing Circuit Using Multi-Winding Transformer. *IEEE Access* **2020**, *8*, 149544–149554. [CrossRef]
21. Jeon, J.; Park, W.; Pyo, S.; Lee, D. Coil design and de-embedding for a novel cell balancing circuit using near-field coupling. In Proceedings of the 2022 37th International Technical Conference on Circuits/Systems, Computers and Communications (ITC-CSCC), Phuket, Thailand, 5–8 July 2022; pp. 546–548.

22. Srinivasan, M.P.; Parimi, A.M. Design of an Isolated Bidirectional Active Cell Balancing circuit for Lithium ion batteries. In Proceedings of the 2022 IEEE 7th International conference for Convergence in Technology (I2CT), Mumbai, India, 7–9 April 2022; pp. 1–6.
23. Conway, T. An Isolated Active Balancing and Monitoring System for Lithium Ion Battery Stacks Utilizing a Single Transformer Per Cell. *IEEE Trans. Power Electron.* **2021**, *36*, 3727–3734. [CrossRef]
24. Kwon, J.-H.; Choi, S.-C.; Zhou, G.; Park, S.-M.; Park, S.-J.; Wang, Y. A Cell-to-cell Voltage Balancing Strategy with Bidirectional Flyback Converter. In Proceedings of the 2024 IEEE 10th International Power Electronics and Motion Control Conference (IPEMC2024-ECCE Asia), Chengdu, China, 17–20 May 2024; pp. 4088–4093.

batteries

MDPI

Article

Development of Modular BMS Topology with Active Cell Balancing

José Gabriel O. Pinto [1,*], João P. D. Miranda [1], Luis A. M. Barros [1] and José A. Afonso [2,*]

[1] ALGORITMI Research Centre/LASI, University of Minho, 4800-058 Guimarães, Portugal; pg47332@alumni.uminho.pt (J.P.D.M.); lbarros@dei.uminho.pt (L.A.M.B.)
[2] Center for Microelectromechanical Systems (CMEMS), University of Minho, 4800-058 Guimarães, Portugal
* Correspondence: gpinto@dei.uminho.pt (J.G.O.P.); jose.afonso@dei.uminho.pt (J.A.A.)

Abstract

This paper presents the design, implementation and experimental validation of a modular battery management system (BMS) featuring active cell balancing. The proposed BMS consists of a master module and multiple slave submodules responsible for monitoring and balancing 22 cells connected in series. The master module collects voltage and temperature data from the slave submodules and measures the battery current to estimate the cells' state of charge (SoC). Each slave module performs cell voltage and temperature measurements and controls a balancing circuit based on dc-dc converters. This work describes in detail the development and validation of the dc-dc converter based in the switched inductor topology, presenting the converter's operational principles, a theoretical and simulation-based analysis of its performance, the implementation of the MOSFETs driver circuits based on PNP transistors and experimental results obtained from a submodule prototype. The results demonstrate the capability of the switched inductor converter to achieve effective voltage equalization by transferring energy from the cells with higher voltages to cells with lower voltages.

Keywords: active cell balancing; battery; battery management system; electric vehicles

1. Introduction

After several decades of limited progress, electric vehicles (EVs) have recently regained significant momentum and are now widely recognized by experts as a key technology for decarbonizing the mobility and transportation sectors [1].

One of the most critical EV components, which currently limits the large-scale adoption of EVs, is the energy storage system. Despite advances in battery research, current technologies rely on scarce and environmentally unfriendly materials, making batteries one of the most expensive components and significantly increasing the overall cost of EVs. In addition to the cost, the lithium-based battery technologies most widely used in EVs are very electrically, thermally, and mechanically sensitive, requiring careful management during operation and storage [2].

EV batteries are usually made up of hundreds of cells connected in series to achieve a voltage high enough to efficiently drive an electric traction motor. Many manufacturers divide the vehicle's battery into submodules with a smaller number of cells. This modularization simplifies handling and improves safety, since each submodule operates at a lower voltage than the total battery pack. Another advantage of this modular approach is fault tolerance: in the event of a cell failure, the submodule can be disabled, allowing the vehicle to continue operating with the remaining submodules [3,4].

To protect the cells mechanically and thermally, the battery enclosure must shield them from mechanical impacts, accommodate volume variations during operation, and ensure proper air or liquid flow for effective heat dissipation. From an electrical point of view, cells are usually assembled with a battery management system (BMS) designed to monitor and protect the cells, thereby preserving their state of health (SoH) [5].

A BMS is an embedded system composed of hardware and firmware that operates in real time and performs several critical functions to preserve the SoH and enhance overall battery performance. One of the most important functions of the BMS is the continuous monitoring of all cell voltages to ensure compliance with the manufacturer's specified operating limits. This includes ensuring that the cell voltage remains above the minimum threshold during discharge and below the maximum threshold during charging. Cells' voltages are also used—alone or in conjunction with other parameters such as instantaneous current—to estimate the state of charge (SoC) [6,7]. Cell temperature is another critical parameter that must be continuously monitored.

Due to manufacturing tolerances, cells of the same model exhibit variations in internal resistance and energy storage capacity. These differences tend to increase over time and usage, requiring an additional BMS function: cell balancing [8].

Cell balancing can be classified as either passive (dissipative) or active (non-dissipative). Passive balancing techniques operate by dissipating the excess of energy of the most charged cells until voltage equalization is achieved across all the cells. The dissipation is typically accomplished by connecting resistors across the cells with the highest voltages. This balancing technique offers simplicity and low cost but suffers from a major drawback: significant energy loss in the form of heat. Conversely, active balancing techniques employ dc-dc power converters to transfer energy from the most charged cells to the discharged ones, achieving higher efficiency at the expense of increased circuit complexity and cost.

Depending on the element used to transfer energy from the most charged cells to the less charged ones, active balancing circuits can be classified as:

- Capacitor-based: These topologies use a capacitor as the intermediate energy storage element, transferring energy from the most charged cells to less charged cells [6,9].
- Inductor-based: These topologies use an inductor as the intermediate energy storage element, transferring energy from the most charged cells to less charged cells [10–12].
- Transformer-based: These topologies use a transformer, which provides magnetic isolation, as the intermediate energy storage element, transferring energy from the most charged cells to less charged cells. In the flyback based topology, energy is first stored in the transformer core and then delivered to the lower-voltage cell [13–17].
- Hybrid: These topologies combine two or more of the aforementioned balancing methods to achieve improved performance in terms of efficiency and response time [9,18,19].

When compared, all balancing circuits have advantages and disadvantages, which makes it difficult to determine a single optimal solution. Some inductor-based topologies offer a good trade-off among component count, circuit complexity and overall performance [6].

In [20], the authors present an analysis and evaluation of a non-dissipative lithium-ion battery balancing method based on a switched inductor topology that transfers energy between series-connected cells to achieve a balanced condition. In [11], a balancing circuit that offers an advantage over conventional designs by enabling the balancing of non-adjacent cells is proposed. In [21], an inductor-based cell equalization technique is investigated. The proposed equalizer demonstrates a faster balancing time compared to other methods, such as switched capacitor and conventional switched inductor methods. A new cell-to-cell active equalization method using planar coupled inductors is proposed and analyzed in [22]. The use of planar core and multi-winding implemented in a printed circuit can be more suitable for industrial applications with low voltage and high current. The overall performance in terms of equalization speed, efficiency, and complexity of implementation is proposed to be studied in a future work. An improved balancing strategy for an inductor-based balancing topology is presented in [23]. The proposed strategy reduces losses and decreases balancing time in comparison to the original strategy.

This paper proposes a modular BMS topology featuring switched inductor active cell balancing. Beyond the design, characterization, and analysis of the switched inductor converter, the paper also describes the development and experimental validation of a full-scale prototype, including the MOSFET driver circuits and the measurement circuits used for cell voltage acquisition.

The remainder of this paper is organized as follows. Section 2 presents the proposed BMS topology and explains its operating principle and functionalities. Section 3 discusses the simulation results of the developed switched inductor converter topology. Section 4 reports the experimental results obtained with both a reduced-scale test setup and a full-scale submodule prototype. Finally, Section 5 concludes the paper with a summary of the main findings.

2. Modular BMS Topology

2.1. Active Versus Passive Balancing Circuits

BMS electronic circuits for cell equalization can be classified as either passive or active balancing circuits. Passive balancing circuits are simple and cheap to implement. However, they dissipate excess energy through resistors connected in parallel with the most charged cells. Active circuits are more complex and costly but enable energy transfer from higher-voltage cells to lower-voltage ones, thereby conserving total battery pack energy while redistributing charge to achieve voltage uniformity among cells.

Figure 1 presents a comparison between passive and active balancing circuits. In Figure 1a, which illustrates active balancing, the main battery current (i_{batt}) and the individual balancing currents ($i_{bat\ x}$) are shown, along with the balancing resistors ($R_{bat\ x}$). These resistors are switched, connected, or disconnected, by the BMS controller depending on each cell's voltage level—activated when balancing is required and disconnected once voltage equalization is achieved. Figure 1b illustrates active balancing, in which dc-dc power converters transfer energy from higher-voltage cells to lower-voltage ones. The main battery current (i_{batt}) and the balancing currents ($i_{bat\ x}$) are also presented. Various dc-dc power converter topologies can be employed for active balancing, each with different trade-offs in terms of efficiency, complexity and cost [6].

Figure 2 illustrates the advantages of employing active balancing circuits over passive balancing approaches. Before balancing, the cells exhibit varying charge levels. After passive balancing, all cells' voltages are equalized; however, the final charge level corresponds to that of the initially least charged cell (e.g., Cell 2). By applying an active balancing circuit, energy is redistributed among cells, allowing all cells to reach the same charge level through controlled energy transfer. Ideally, neglecting conversion losses in the dc-dc converters, the total energy of the battery pack remains unchanged after active balancing.

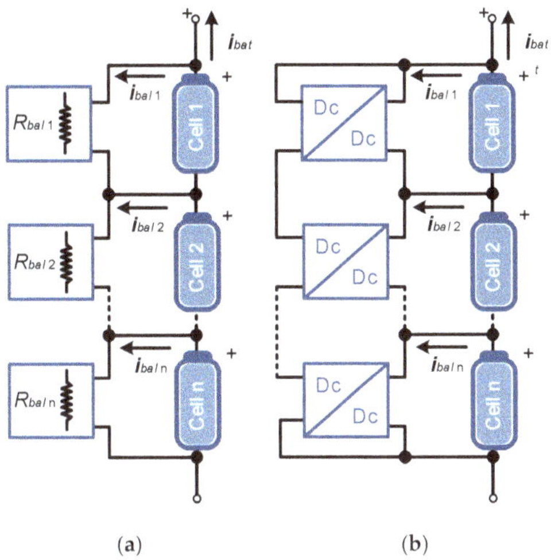

Figure 1. Types of balancing circuits: (**a**) passive cell balancing; (**b**) active cell balancing.

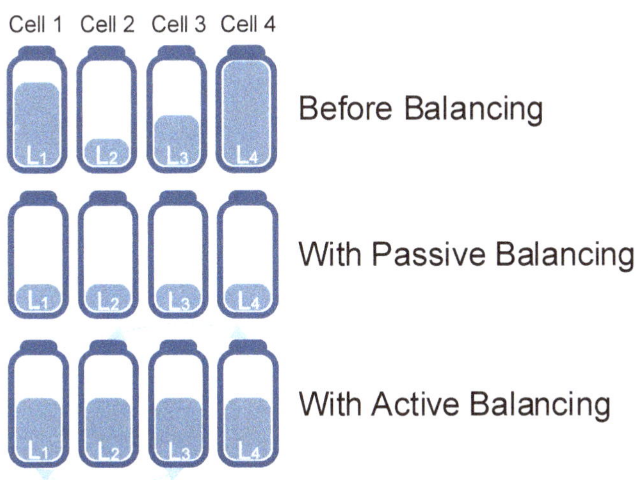

Figure 2. Benefits of active balancing circuits.

2.2. Proposed BMS Architecture

This work proposes a modular BMS architecture designed to monitor and manage battery packs containing a large number of cells organized into various submodules. The architecture consists of a master device and a configurable number of slave devices, one for each battery submodule.

Each slave incorporates a dedicated battery monitor circuit, the BQ76940 model from Texas Instruments, Dallas, TX, USA, which measures cells' voltage and temperature via an I^2C communication interface. Each slave device also includes a microcontroller responsible for acquiring cell voltage and temperatures from the BQ76940, communicating with the master device through an isolated controller area network (CAN) interface, and generating pulse width modulation (PWM) control signals to the balancing circuits. Each slave was designed for a maximum of 22 cells connected in series.

Figure 3 presents a block diagram of the proposed modular BMS architecture with active cell balancing functionality. Each slave board is directly mounted onto its corresponding battery submodule and connects to a CAN bus, which also supplies 12 V to the galvanically isolated CAN transceiver integrated into the slave board.

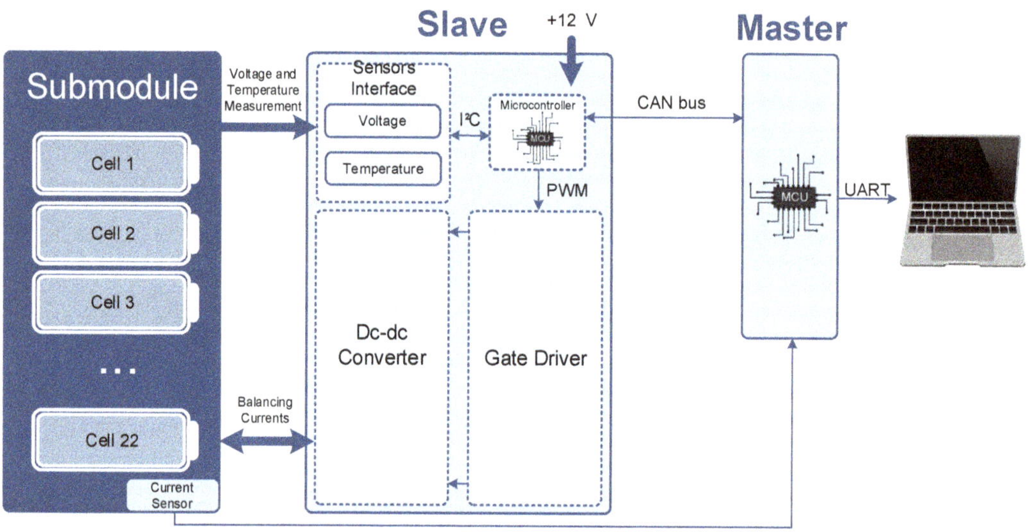

Figure 3. Block diagram of the proposed BMS modular architecture with active cell balancing.

Although the proposed BMS is a complex system comprising both hardware and firmware, this work focuses primarily on the hardware design, specifically the dc-dc converters used for active cell balancing. For this application, these converters must be designed to achieve high efficiency while remaining compact and cost-effective.

2.3. Dc-Dc Converter for Active Cell Balancing

The dc-dc converter for active cell balancing in the proposed BMS is based on the switched inductor topology. To balance a battery pack with n cells, this configuration requires a total of $2(n-1)$ fully controlled semiconductors and $n-1$ inductors.

Figure 4 illustrates the switched inductor topology implemented in the proposed BMS and its operation. Figure 4a presents a simplified schematic of the converter employed for the active cell balancing. For analysis purposes, a configuration with three cells is considered. In this case, the circuit includes four controlled semiconductors, S_1, S_2, S_3, and S_4; two inductors, L_1 and L_2; and one current flowing in each inductor, i_{L1} and i_{L2}.

The converter's operation consists of two distinct stages. In the first stage, switches S_1 and S_2 are turned on, allowing energy transfer from the higher-voltage cells to the inductors, while switches S_3 and S_4 are off, as shown in Figure 4b. In the second stage, S_1 and S_2 are turned off, while S_3 and S_4 are turned on, enabling the stored energy in the inductors to be delivered to the adjacent lower-voltage cells, as illustrated in Figure 4c.

Considering the case in which Cell 1 has a higher voltage than Cell 2, switch S_1 is turned on while switch S_3 remains off, as illustrated in Figure 5a. When switch S_1 is turned off, charging Cell 2, as shown in Figure 5b, current flows through the freewheeling diode of switch S_3. This switching sequence occurs at high frequency—on the order of tens of kilohertz—allowing a gradual transfer of energy from Cell 1 to Cell 2.

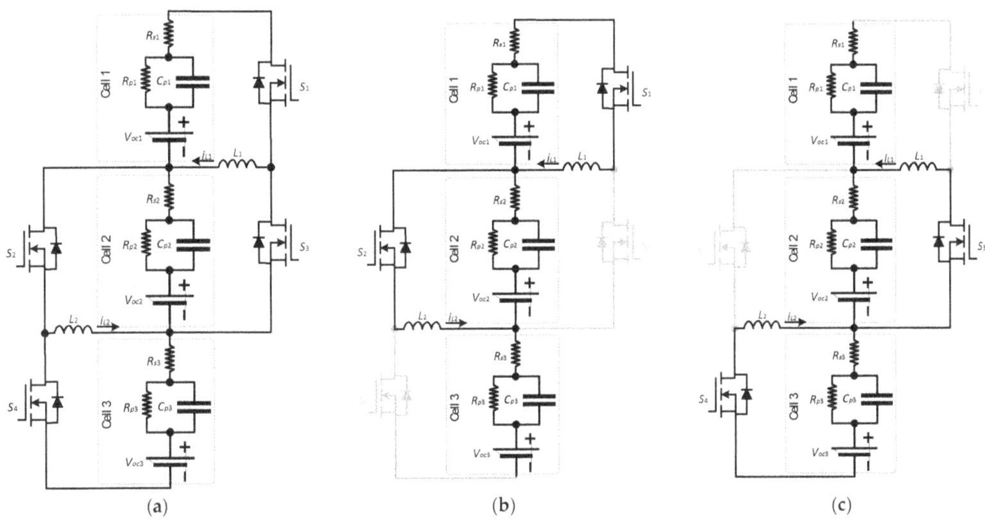

Figure 4. Switched inductor converter used in the proposed BMS for active cell balancing: (**a**) Simplified circuit; (**b**) Stage 1 with S_1 and S_2 turned on, S_3 and S_4 turned off; (**c**) Stage 2 with S_3 and S_4 turned on, S_1 and S_2 turned off.

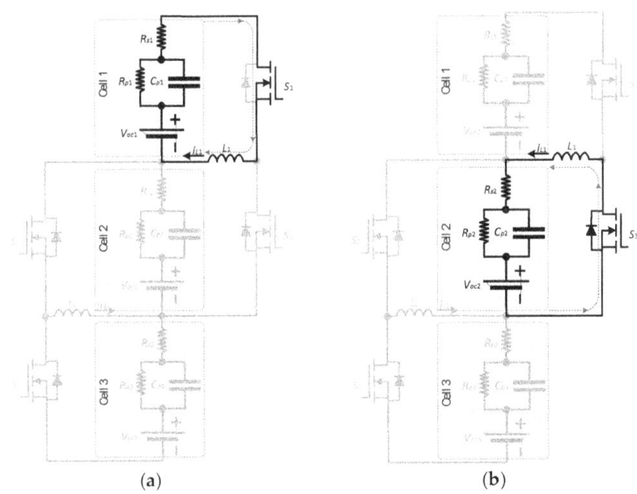

Figure 5. Current path when the Cell 1 voltage is higher than Cell 2 voltage: (**a**) S_1 is turned on; there is an energy transfer from Cell 1 to inductor L_1; (**b**) S_1 is turned off and S_3 is turned on; the energy stored in L_1 follows to Cell 2.

3. Simulation of the Switched Inductor Converter

3.1. Cell Model

To simulate the proposed BMS topology, a Thevenin equivalent model of a battery cell was developed [24]. The cell model consists of a series resistor representing the internal resistance of the cell and a parallel resistor-capacitor set modeling the cell capacity and self-discharge behavior. An ideal voltage source corresponding to the cut-off cell voltage is also included. The cell used in this study was the Samsung, Suwon, Sud Corea, INR21700 40T. The parameters of the equivalent circuit, such as the series resistance and capacitance, were determined based on the manufacturer's datasheet. Table 1 summarizes the main cell specifications of the selected cell.

Table 1. Specifications from the Samsung cell model INR21700 40T.

Parameter	Value
Nominal voltage	3.6 V
Maximum voltage	4.2 V
Discharge cut-off voltage	2.5 V
Internal Resistance (1 kHz)	12 mΩ
Capacity	4 Ah
Max. continuous discharge	35 A
Cell weight	67 g
Cell dimensions	21 × 70 mm

Based on the data presented in Table 1, the maximum and minimum cell voltage were assumed to be 4.2 V and 2.5 V, respectively. The internal resistance R_s was considered to be 12 mΩ. According to Equation (1), the value of the capacitor (C_p) used in the computer simulation can be obtained from the ratio between the cell charge Q (Ampere-hours) and the corresponding voltage variation ΔV at the cell terminals. By substituting the values into Equation (1), the capacitance value for the equivalent cell model can be determined. Due to the large value of C_p, the simulation of the equivalent electrical model would require long computation times. To overcome this, a time-scaling factor of 1 h to 1 s was adopted in the simulation. The adjusted value of C_p considered for the equivalent model was 2.35 F. The resistor R_p, representing the cell self-discharge process—a phenomenon that does not manifest significantly over short time intervals—was set to 1 MΩ.

$$C_p = \frac{Q}{\Delta V} \tag{1}$$

3.2. Reduced-Scale Simulation Model

Figure 6 shows the schematic of the switched inductor balancing topology in the PSIM 11.0 simulator. The semiconductor switches are driven by two 25 kHz PWM signals with an 180° phase-shift and a 50% duty-cycle. *Vcell1*, *Vcell2*, and *Vcell3* correspond to the voltages in Cells 1, 2, and 3, respectively. To analyze the converter's behavior, simulations were carried out under three different operating conditions. In the first test, according to the circuit shown in Figure 6, three cells and two 25 μH inductors were used. Series resistors connected to the MOSFETs were included to emulate the on-state resistance $R_{DS(ON)}$. A value of 30 mΩ was adopted.

3.3. Reduced-Scale Simulation Results

To evaluate the circuit's balancing performance, the simulation was initialized under a highly unbalanced condition with *Vcell1* = 3 V, *Vcell2* = 3.5 V, and *Vcell3* = 4 V. As shown in Figure 7a, all three voltages converged toward an equalized value of approximately 3.5 V. The balancing time was less than 2 s, which corresponds to less than 2 h under real operating conditions, and the resulting voltage ripple is minimal. Figure 7b provides a detailed view of the cell voltages, revealing a small ripple that is more pronounced in Cell 2. This occurs because Cell 2 is simultaneously receiving energy from Cell 3 and transferring energy to Cell 1.

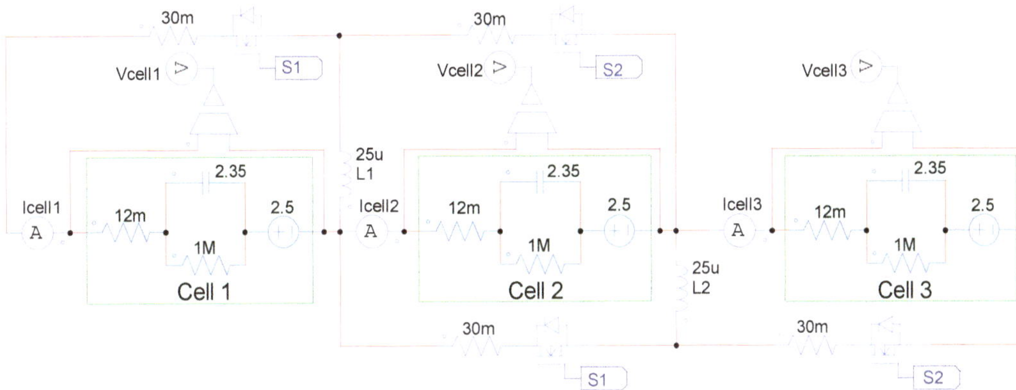

Figure 6. Simulation model of the switching inductor converter implemented in the PSIM 11.0 simulation software.

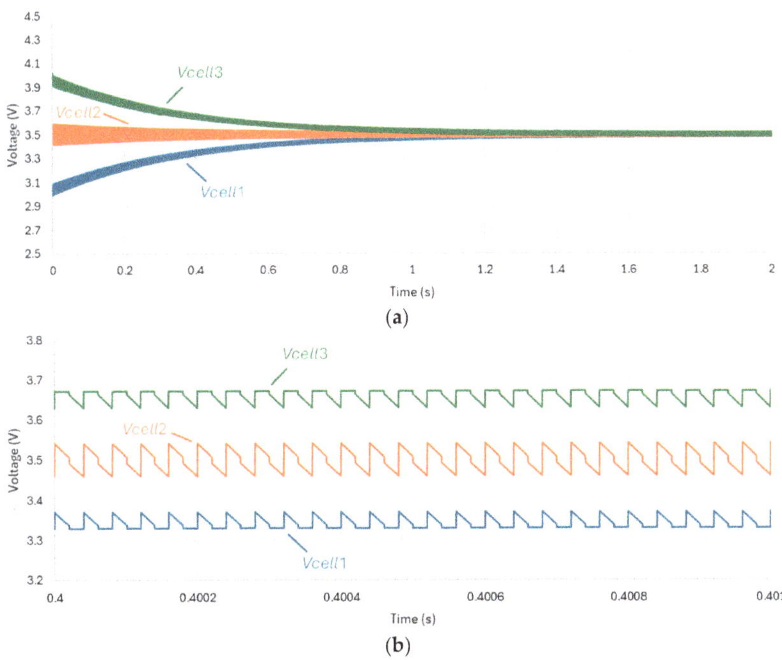

Figure 7. Simulation results of the switched inductor converter for initial cell voltages *Vcell*1 = 3 V; *Vcell*2 = 3.5 V and *Vcell*3 = 4 V: (**a**) Voltage in the cells during the balancing period; (**b**) Detail of the cells ripple voltages.

Regarding the waveform of the inductor currents i_{L1} and i_{L2} illustrated in Figure 8a, it can be observed that since Cell 1 initially has a lower voltage, inductor *L*1 exhibits a positive average current. In contrast, inductor *L*2 presents a negative average current, as Cell 3 operates as the energy source. Once the balancing process is completed, the average currents in both inductors become zero, indicating that there is no longer effective energy transfer between the cells. At this stage, the MOSFETs must cease switching to minimize unnecessary power losses. Figure 8b provides a detailed view of the inductor currents, highlighting their characteristic triangular waveforms.

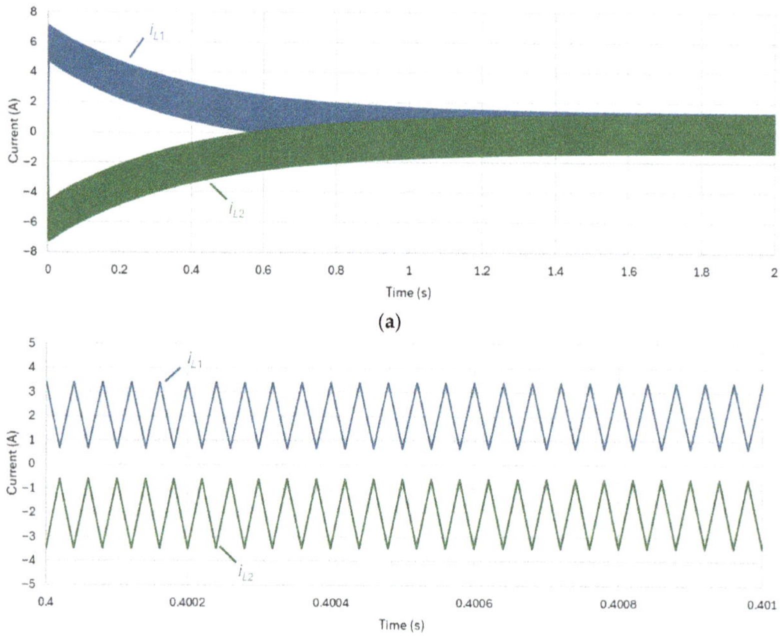

(b)

Figure 8. Simulation results of the switched inductor converter for initial cell voltages $Vcell1$ = 3 V, $Vcell2$ = 3.5 V, and $Vcell3$ = 4 V: (**a**) Currents in the inductors $L1$ and $L2$ during the balancing period; (**b**) Detail of the inductor currents.

Analyzing the cells' currents as illustrated in Figure 9a, it can be observed that the currents i_{cell1} and i_{cell3} exhibit a decreasing trend over time. This behavior corresponds to the gradual convergence of the cell voltages toward their average value. The average current in Cell 1, initially positive, and in Cell 3, initially negative, both approach zero as voltage equalization is achieved. Figure 9b provides a detailed view of the cell currents. The current in Cell 1 (i_{cell1}) exhibits a negative average value, indicating that the cell is absorbing energy, since the current direction is defined as positive when the cell delivers energy to the circuit. Conversely, the current in Cell 3 (i_{cell3}) exhibits a positive average value, confirming that it is receiving energy during the balancing process.

From the instantaneous power waveforms shown in Figure 10, it can be observed that Cell 1, the least charged, exhibits a negative average power during the balancing period, indicating that it is absorbing energy. Conversely, Cell 3, the most charged, presents a positive average value, confirming that it is supplying energy to the system. The average power in Cell 2 remains approximately zero throughout the balancing process, as it alternately receives energy from Cell 3 and transfers it to Cell 1 during successive switching stages. After voltage equalization is achieved, the average power of all cells converges to zero, indicating that the energy flow among cells has ceased.

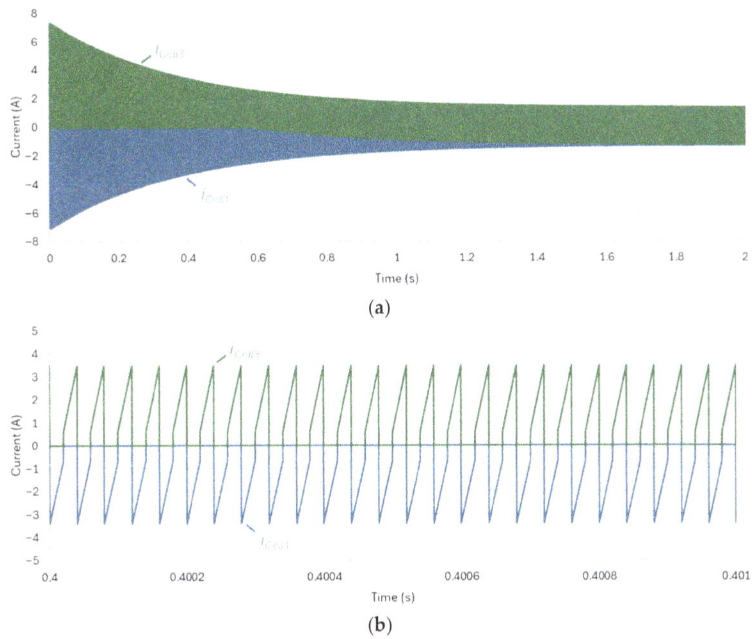

Figure 9. Simulation results of the switched inductor converter for initial cell voltages *Vcell*1 = 3 V, *Vcell*2 = 3.5 V, and *Vcell*3 = 4 V: (**a**) Currents in Cell 1 and Cell 3 during the balancing period; (**b**) Detail of the cell's currents.

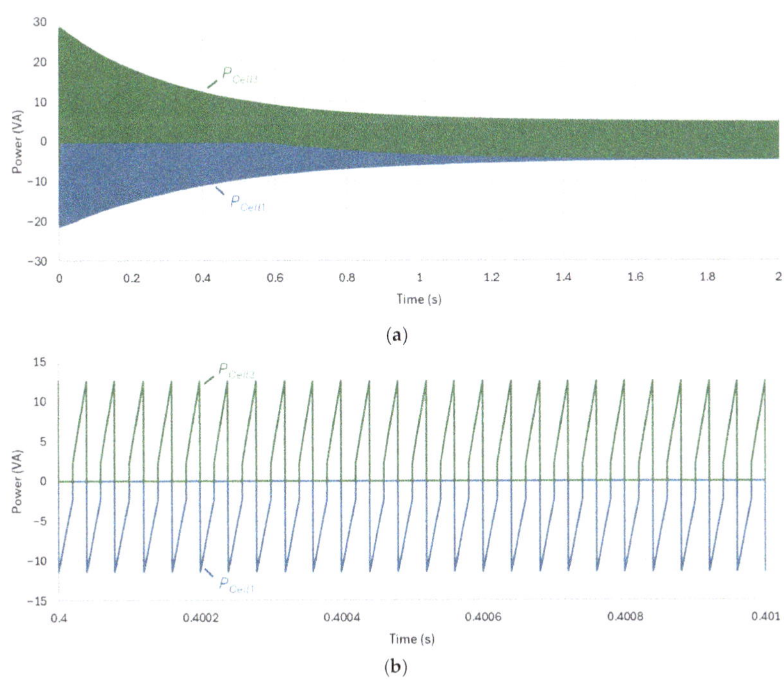

Figure 10. Simulation results of the switched inductor converter for initial cell voltages *Vcell*1 = 3 V, *Vcell*2 = 3.5 V, and *Vcell*3 = 4 V: (**a**) Power in Cell 1 and Cell 3 during the balancing period; (**b**) Detail of the cell's power.

3.4. Full-Scale Simulation Model and Results

After validating the switched inductor converter using a reduced number of cells, a full-scale simulation model was developed to represent the 22-cell configuration of the BMS submodule. Figure 11 presents the complete simulation model of the switching inductor converter implemented in the PSIM 11.0 software.

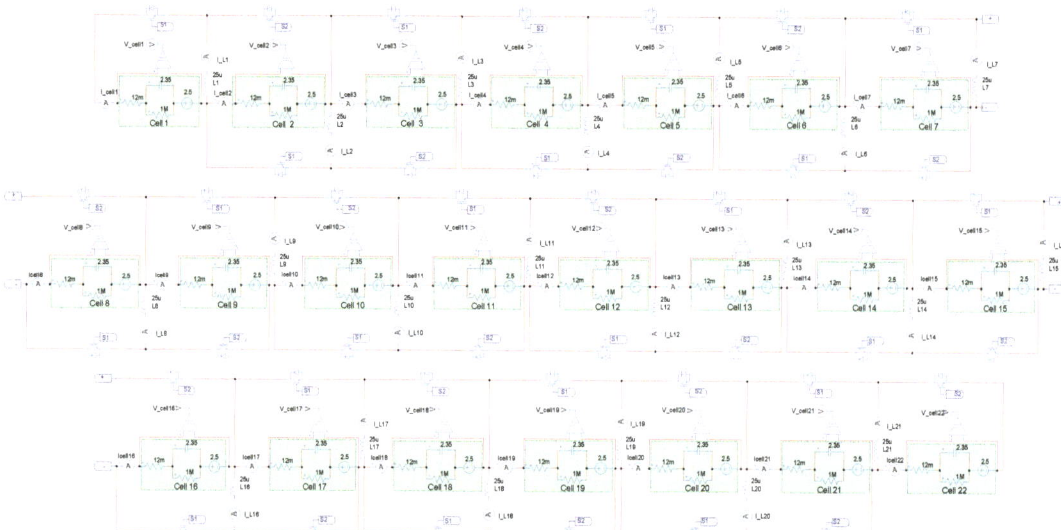

Figure 11. Complete simulation model of the switching inductor converter implemented in the PSIM 11.0 simulation software.

Figure 12 shows the simulation results of the previous simulation model, where the voltages of all 22 cell voltages converge toward the same value. In this case, a smaller imbalance level was considered, with the most discharged cell at 3.3 V and the most charged cell at 3.7 V. It can also be observed that, for the initial conditions considered, the balancing time is less than 1 s, corresponding to less than 1 h in real operation. These results validate the ability of the switched inductor converter to actively balance the 22 cells of the BMS submodule.

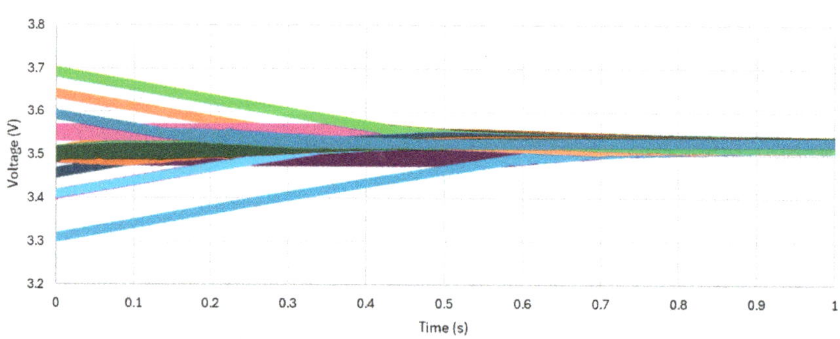

Figure 12. Simulation results of the switched inductor full-scale model: Voltages of the 22 cells of the BMS submodule during the balancing process.

4. Experimental Validation of the Switched Inductor Converter

4.1. MOSFETs Driver Circuit

Since the battery module cells are connected in series and the source pins of the MOSFETs are not referenced to the ground, a dedicated driving circuit is required to control the MOSFETs without creating short-circuits between cells. One possible solution is to use isolated gate drivers. Although several integrated circuits from different manufacturers could meet the system requirements, their use would increase the overall cost. Therefore, a simple and low-cost driver circuit was specifically developed for this application.

Figure 13 shows the developed driver circuit for the switched inductor MOSFETs, all of N-type, considering only two cells for simplicity. The circuit consists of two main NPN transistors, *Tn1* and *Tn2*, responsible for biasing the PNP transistors, one for each MOSFET. When the PWM signal is high, the NPN transistors start to conduct, allowing current to flow between the emitter and base terminals of the PNP transistors and forcing them to start conducting. The base-emitter current is sufficiently high to drive the PNP transistors into saturation, minimizing the emitter-collector voltage and ensuring that the corresponding MOSFET operates in the triode region with very low conduction losses.

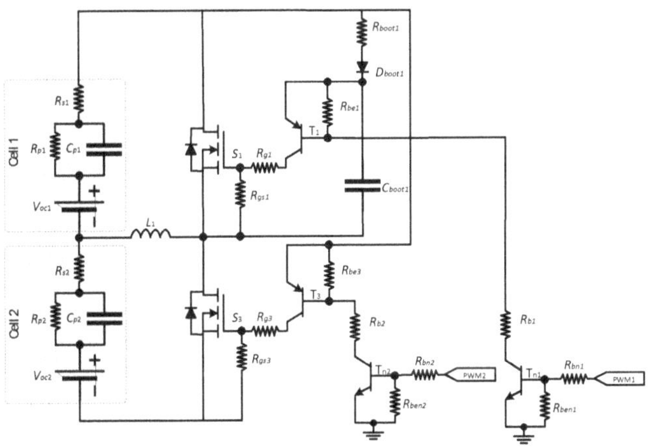

Figure 13. PNP transistor-based circuit driver for the switched inductor converter MOSFETs.

To guarantee operation in the triode region, the emitter of each PNP transistor is connected to the positive terminal of the next cell, ensuring that the gate-source voltage (V_{GS}) equals the sum of the voltage of the two adjacent cells. This configuration maintains V_{GS} sufficiently higher than the MOSFET threshold voltage (V_T). It is important to note that if only a single cell whose voltage could drop to 2.5 V was used, this would approach V_T, and the MOSFETs could conduct outside the triode region, increasing the losses.

For the topmost cell in the series, where no subsequent cell is available, a bootstrap circuit consisting of diode *Dboot*1, the capacitor *Cboot*1, and the resistor *Rboot*1 was employed. Each time MOSFET S_3 is turned on, capacitor *Cboot1* charges through the diode and resistor. Since S_3 switches at 25 kHz, the capacitor voltage quickly reaches a steady-state value equal to the sum of voltages of Cell 1 and Cell 2, providing an adequate voltage to drive MOSFET S_1 through transistor T_1.

4.2. Switched Inductor Converter Experimental Results

This section presents the experimental results obtained with the BMS prototype developed using the switched inductor converter topology. For the initial test validation, a prototype BMS board was designed to operate with six cells. To measure the currents, low-value shunt resistors were inserted at selected measurement points, namely in series with the cells, inductors, and MOSFETs. To prevent excessive current in the case of circuit malfunction, 1 Ω resistors were temporarily connected in series with the inductors; these resistors were later bypassed using wire conductors for the final tests. A four-channel Tektronix, Beaverton, OR, USA, TPS 2024B oscilloscope was employed to acquire the experimental waveforms. Figure 14 shows the developed 6-cell switched inductor converter prototype for the experimental validation of the BMS active cell balancing circuit.

(a) (b)

Figure 14. Prototype board of the 6-cell switched inductor converter for the BMS active cell balancing circuit: (**a**) Top view of the prototype board; (**b**) Bottom view of the prototype board.

The initial tests were performed to validate the proper operation of the MOSFET drive circuits. Figure 15a presents the PWM signals, including a 3 μs dead time, and the base-emitter voltages of the two NPN transistors. Figure 15b shows the NPN base-emitter versus the collector-emitter voltages, demonstrating that the propagation delay is minimal and the dead time is adequately maintained.

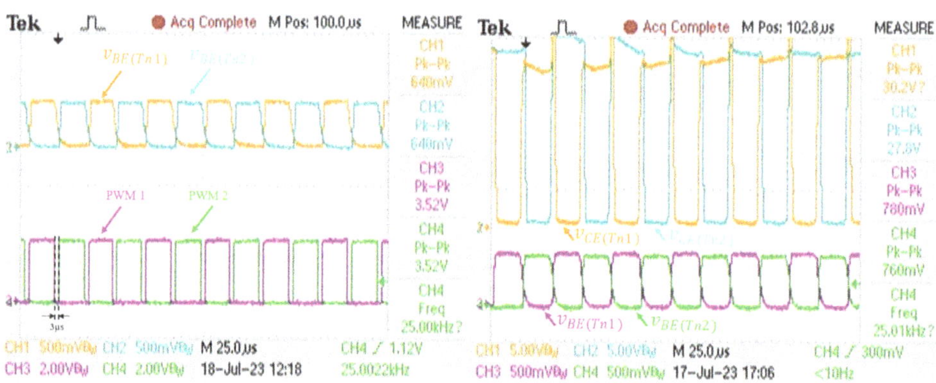

(a) (b)

Figure 15. Experimental results of the driver circuits: (**a**) PWM signals (PWM1, PWM2) and base-emitter voltages of the NPN transistors ($v_{BE(\mathrm{Tn1})}$, $v_{BE(\mathrm{Tn2})}$); (**b**) Base-emitter ($v_{BE(\mathrm{Tn1})}$, $v_{BE(\mathrm{Tn2})}$) and collector-emitter ($v_{CE(\mathrm{Tn1})}$, $v_{CE(\mathrm{Tn2})}$) voltages of the NPN transistors.

Figure 16a shows the emitter-base voltages of the PNP transistors T_1 and T_3 along with their respective emitter-collector voltages at circuit startup. The progressive increase in the emitter-collector voltage of transistor T_1 corresponds to the charging of the bootstrap capacitor. Figure 16b presents the same signals in a steady state, showing that the emitter-collector voltage of transistor T_1 stabilizes after a few switching cycles. When the voltage

$v_{EC(T1)}$ is zero, MOSFET S_1 conducts; when $v_{EC(T3)}$ is zero, the MOSFET S_3 conducts. The dead time remains essentially constant throughout the operation.

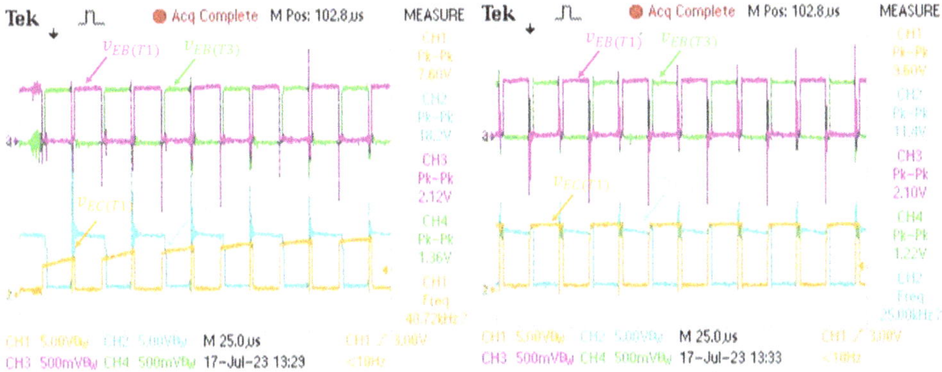

(a) (b)

Figure 16. Experimental results of the driver circuits: (**a**) Emitter-base ($v_{EB(T1)}$, $v_{EB(T2)}$) and emitter-collector ($v_{EC(T1)}$, $v_{EC(T2)}$) voltages of the PNP driver transistors, at circuit startup; (**b**) Emitter-base ($v_{EB(T1)}$, $v_{EB(T2)}$) and emitter-collector ($v_{EC(T1)}$, $v_{EC(T2)}$) voltages of the PNP driver transistors, in steady state operation.

Figure 17a shows the V_{GS} voltages of MOSFETs S_1 and S_3 at circuit startup, when the bootstrap capacitor begins charging. Figure 17b shows the same signals in a steady state. The V_{GS} waveforms maintain a dead time between them, with v_{GS1} reaching 8.6 V and v_{GS3} reaching 8.4 V, values that are sufficient to ensure that the MOSFETs operate in the triode region.

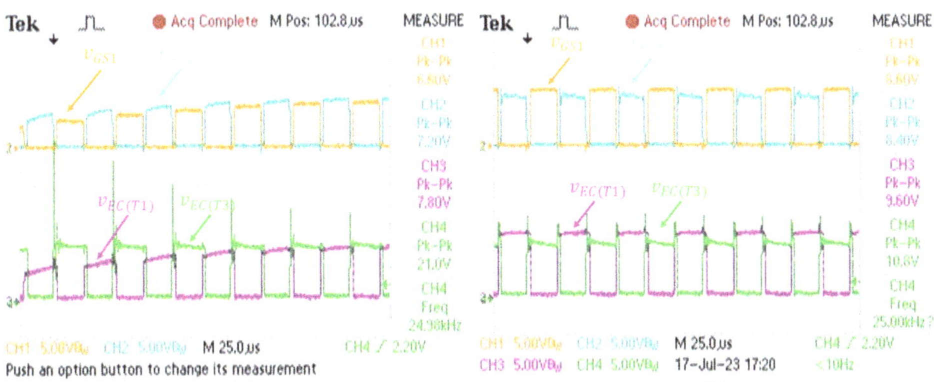

(a) (b)

Figure 17. Experimental results of the driver circuits: (**a**) Gate-source (v_{GS1}, v_{GS2}) voltages of the MOSFETs and emitter-collector ($v_{EC(T1)}$, $v_{EC(T2)}$) voltages of the PNP driver transistors, at circuit startup; (**b**) Gate-source (v_{GS1}, v_{GS2}) voltages of the MOSFETs and emitter-collector ($v_{EC(T1)}$, $v_{EC(T2)}$) voltages of the PNP driver transistors, in steady state operation.

Figure 18 shows in detail the overall delay propagation time of the PNP gate driver circuit. Figure 18a shows that the delay between the PWM signal of the microcontroller (PWM) and the gate-source voltage of the MOSFET (v_{GS}) is about 870 ns during turn-on. Figure 18b shows that the delay between the PWM signal of the microcontroller (PWM) and the gate-source voltage of the MOSFET (v_{GS}) is about 1.18 µs during turn-off. As expected, the turn-on is faster than turn-off, justifying the use of a dead time between the PWM channels.

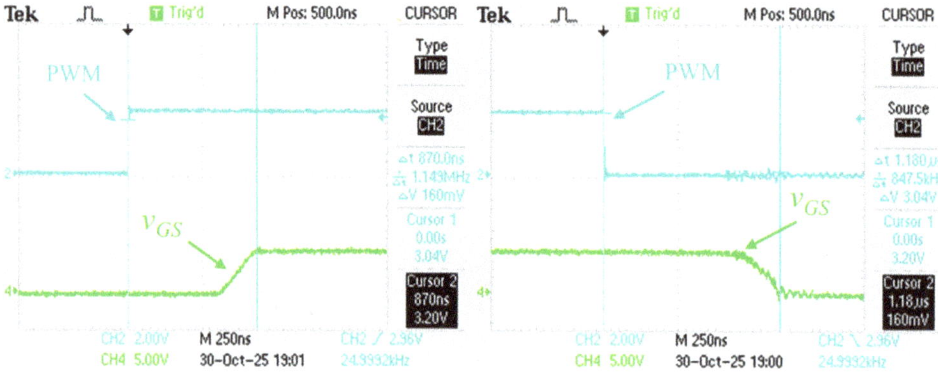

(a) (b)

Figure 18. Detail of delay time between the PWM signal of the microcontroller (PWM) and the gate-source voltage of the MOSFET (v_{GS}) with the PNP gate driver circuit: (**a**) During turn-on; (**b**) During turn-off.

Figure 19 presents the switched inductor experimental results for a condition in which Cell 1 has a voltage of 3.4 V and all the other cell voltages are 3.92 V. Figure 19a shows the currents through MOSFETs S_1 and S_3 along with the emitter-collector voltages of the PNP driver transistors ($v_{ec(T1)}$ and $v_{ec(T3)}$). Figure 19b presents the current in inductor 1 (i_{L1}), the current in Cell 1 (i_{cell1}), and the emitter-collector voltages of the PNP driver transistors ($v_{ec(T1)}$ and $v_{ec(T3)}$). The average current in Cell 1 is negative, indicating that it is being charged, as expected, since its voltage is lower than that of the other cells.

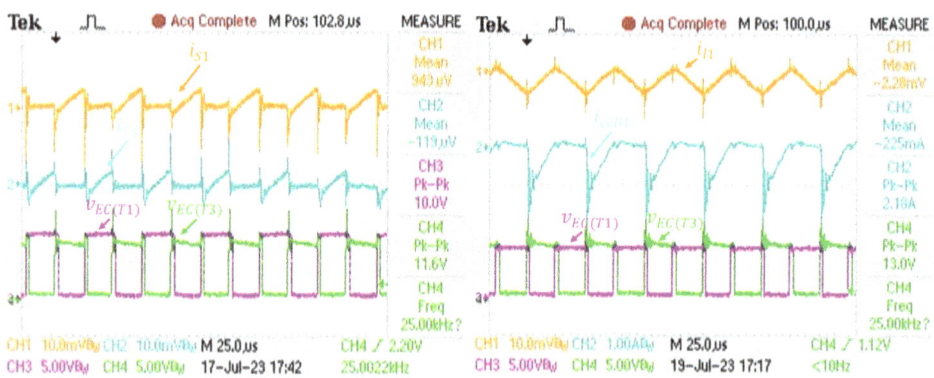

(a) (b)

Figure 19. Experimental results of the switched inductor circuit for the condition v_{cell1} = 3.40 V and $v_{cell2} = v_{cell3} = v_{cell4} = v_{cell5} = v_{cell6}$ = 3.92 V: (**a**) current in the MOSFET S_1, current in the MOSFET S_3, and emitter-collector voltage of the PNP driver transistors ($v_{ec(T1)}$ and $v_{ec(T3)}$); (**b**) current in the inductor 1 (i_{L1}), current in cell 1 (i_{cell1}) and emitter-collector voltage of the PNP driver transistors ($v_{ec(T1)}$ and $v_{ec(T3)}$).

The balancing current of the switch inductor converter depends on the cell's voltage difference and on the resistances in the path, that is, the internal resistance of the inductor, MOSFET R_{DS_on}, the resistance of the copper trails, the internal resistance of the cells, etc.

To assess the maximum balancing current with the developed converter, a scenario with Cell 1 and Cell 3 near the maximum voltage and Cell 2 at a considerably lower voltage was used. In this scenario, both Cell 1 and Cell 3 contribute to charging Cell 2. Figure 20 shows the achieved results. Figure 20a shows the voltage (v_{cell1}) and current (i_{cell1}) in Cell 1 together with the voltage (v_{cell2}) and current (i_{cell2}) in Cell 2. Figure 20b shows the voltage (v_{cell3}) and current (i_{cell3}) in Cell 3 together with the voltage (v_{cell2}) and current (i_{cell2}) in Cell 2. The voltage and current in Cell 2 are common to the two figures. As shown in the figures, Cell 2 is charging with an average current of 908 mA and a peak value of 1.44 A.

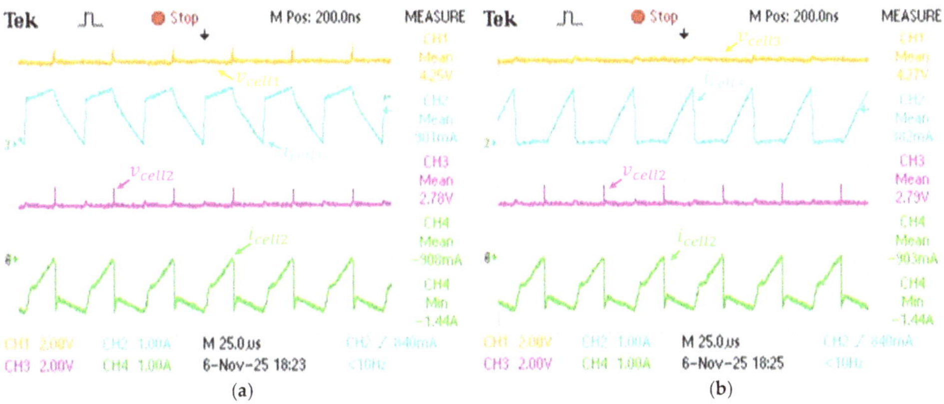

Figure 20. Experimental results of the switched inductor circuit for the condition v_{cell1} = 4.25 V, v_{cell2} = 2.78 V, and v_{cell3} = 4.27 V: (**a**) Voltage in Cell 1 (v_{cell1}) and current in Cell 1 (i_{cell1}); voltage in Cell 2 (v_{cell2}) and current in Cell 2 (i_{cell2}); (**b**) Voltage in Cell 3 (v_{cell3}) and current in Cell 3 (i_{cell3}); voltage in Cell 2 (v_{cell2}) and current in Cell 2 (i_{cell2}).

To assess the energy transfer efficiency, a circuit with only two unbalanced cells was used, and the active power (average value of the instantaneous power) was measured in the two cells. For this procedure, the MATH channel of the oscilloscope was used to calculate the product of the voltage by the current in the cell, resulting in the instantaneous power. Finally, the average value of the MATH channel was calculated to obtain the active power. The efficiency was then calculated by dividing the active power of the cell with lower voltage (Cell 2 that receives energy) by the active power of the cell with higher voltage (Cell 1 that provides energy). This methodology allows us to assess the energy transfer efficiency with high bandwidth and high sampling rate equipment.

Figure 21 shows results of the energy transfer efficiency measurements. Figure 21a shows the voltage (v_{cell1}), current (i_{cell1}) and instantaneous power (p_{cell1}) waveforms in Cell 1. The average power is positive, meaning that the cell is providing energy. Figure 21b shows the voltage (v_{cell2}), current (i_{cell2}) and instantaneous power (p_{cell2}) waveforms in Cell 2. The average power in this case is negative, meaning that the Cell 2 is receiving energy. By dividing the average power of Cell 2 by the average power of Cell 1, an energy transfer efficiency of 84% was obtained.

The experimental results obtained with this small-scale prototype are aligned with the expectations motivating the development of a full-scale 22-cell prototype.

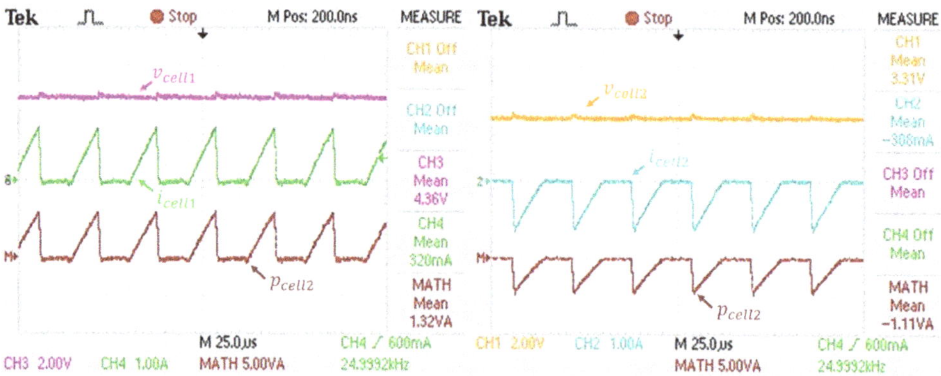

(a) (b)

Figure 21. Experimental results of the switched inductor circuit for the condition v_{cell1} = 4.36 V and v_{cell2} = 3.31 V: (**a**) Voltage in Cell 1 (v_{cell1}), current in Cell 1 (i_{cell1}), and instantaneous power in Cell 1 (p_{cell1}).; (**b**) Voltage in Cell 2 (v_{cell2}), current in Cell 2 (i_{cell2}), and instantaneous power in Cell 2 (p_{cell2}).

4.3. Final Full-Scale Prototype

After validating the switched inductor converter and confirming the correct operation of the driver circuit, a full-scale prototype was developed with the goal of creating a compact PCB. The final prototype, shown in Figure 22, does not include the shunt resistors for current measurements or the 1 Ω limiting resistors in series with inductors used during initial validation. Figure 22a shows the top of the PCB, divided into five functional areas: (i) Area 1 houses the switched inductor converter, including the MOSFETs and inductors; (ii) Area 2 contains the driver circuit; (iii) Area 3 comprises the circuits for measuring cell voltages and temperatures; (iv) Area 4 hosts the microcontroller responsible for acquiring the cells' voltages and temperatures, communicating with the master module and controlling the cell balancing via PWM signals; and (v) Area 5 contains the power and the CAN communication interface connector. Figure 22b shows the bottom of the PCB, containing the remaining components of the switched inductor converter and gate driver circuit.

After assembling the full-scale prototype board, initial tests were performed to verify correct operation of the circuits. The prototype was then connected to the cells and a balancing test was conducted.

Figure 23 shows the experimental results obtained with the full-scale BMS prototype, showing the voltages of three consecutive cells during 2 h and 30 min of balancing operation. The initial and final cell voltages are presented in Table 2 with the estimated initial and final SoC. During balancing, the voltages of Cell 1 and Cell 3 decreased, while the voltages of Cell 2 and Cell 3 increased, as expected, since higher-voltage cells transfer energy to lower-voltage cells. The BMS reduced the initial voltage difference between Cell 1 and Cell 4 from 1.18 V to approximately 0.47 V and the initial SoC difference from 91.2% to 49.4%. The results were achieved using a digital oscilloscope with four non-isolated channels combined with differential voltage probes to allow the measurement of individual cell voltages.

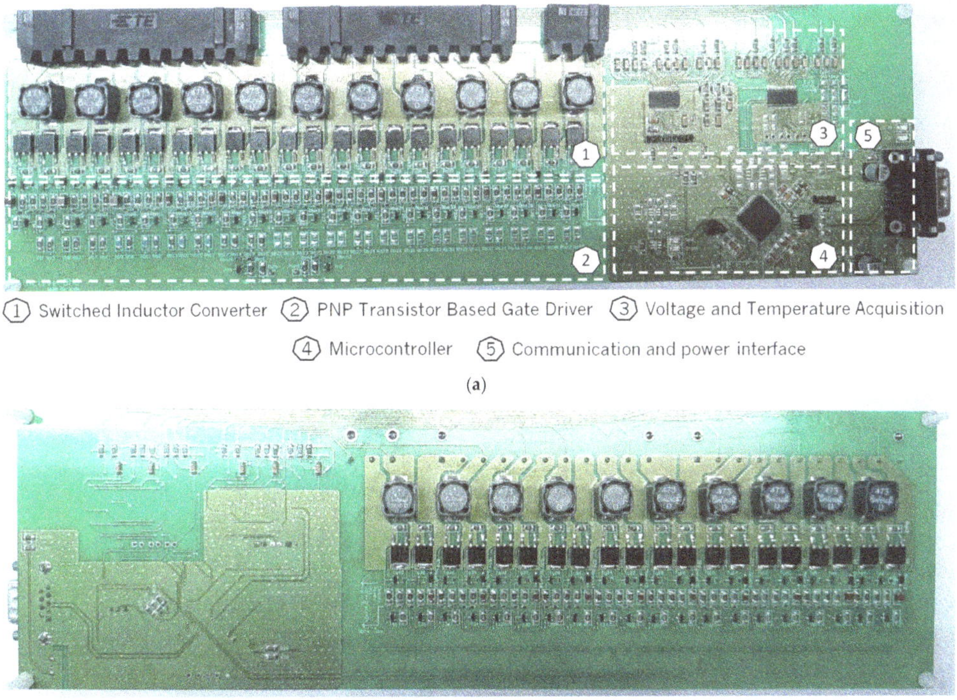

(a)

(b)

Figure 22. Final full-scale prototype for a 22-cell BMS submodule: (**a**) Top view of final BMS PCB; (**b**) bottom view of the BMS PCB.

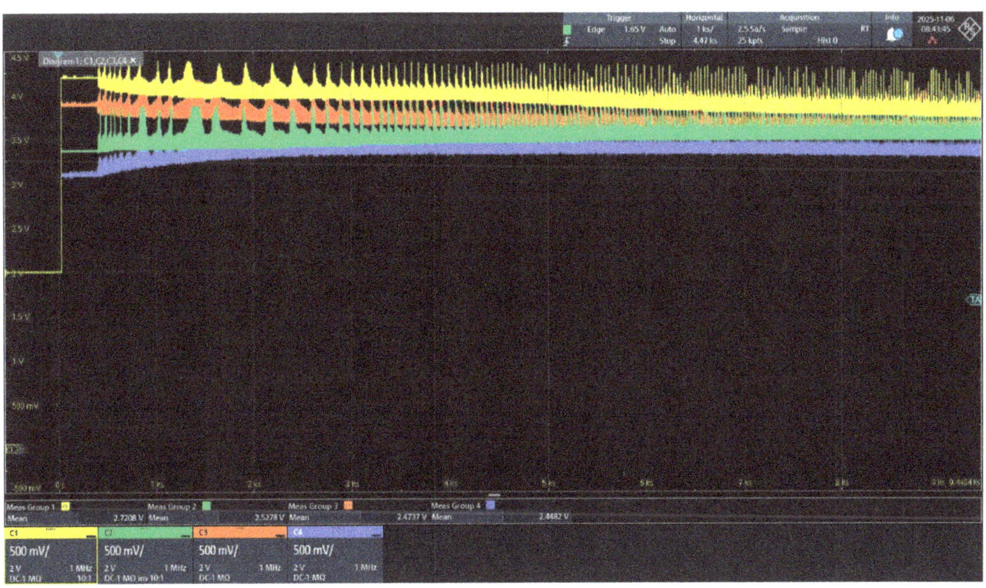

Figure 23. Full-scale BMS prototype experimental results: Voltages of four consecutive cells during 2 h and 30 min balancing operation.

Table 2. Summary results of the 2 h and 30 min balancing operation.

Cell	Initial Voltage	Final Voltage	Initial SoC	Final SoC
Cell1	4.19 V	3.83 V	100%	76.2%
Cell2	3.36 V	3.55 V	27.4%	46.9%
Cell3	3.89 V	3.62 V	87.7%	53.9%
Cell4	3.01 V	3.36 V	8.8%	26.8%

During the cell equalization process, a thermal analysis of the BMS circuit was done. As shown in Figure 24, the thermal analyses reveal that a maximum temperature of 73.3 °C was measured in one of the switched inductor converter MOSFETs.

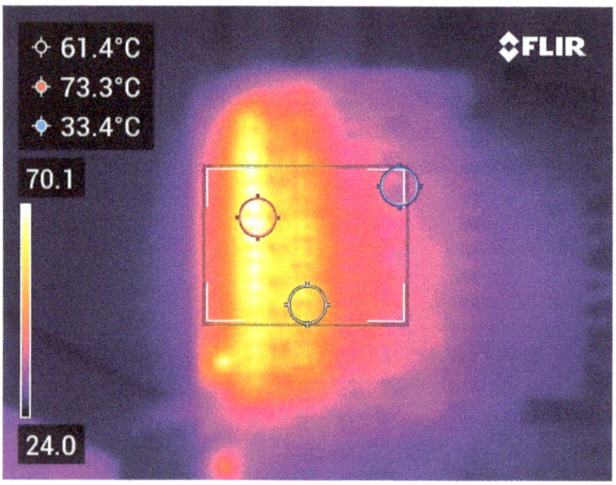

Figure 24. Full-scale BMS prototype experimental results: Temperature of the switched inductor circuit components during a long-time balancing operation.

5. Conclusions

This paper presents the development of modular BMS topology with active cell balancing. The BMS consists of a master module and a configurable number of slave modules that communicate via an isolated CAN bus. The focus of the work was on the switched inductor dc-dc circuit used for active voltage balance.

The topology was first studied and validated through simulations in PSIM 11.0 software, employing a Thevenin equivalent battery model adapted to a time scale of 1 h per 1 s of simulation, i.e., 1 s of simulation time corresponds to 1 h in a real setup.

Following a successful simulation validation, a reduced-scale prototype was developed to experimentally validate the switching inductor converter and the MOSFET driver circuits, based on PNP and NPN bipolar transistors especially designed for this topology. The experimental results with the reduced-scale prototype were consistent with expectations.

A full-scale prototype was subsequently developed for a 22-cell battery submodule connected in series.

Experimental tests of the prototype revealed an energy transfer efficiency of 84% and a balancing current of 908 mA. However, these results are highly dependent on the cells' voltage differences.

The prototype successfully balanced cells from an initial voltage difference of 1.18 V to 0.47 V in 2 h and 30 min. The developed BMS will undergo further electrical and mechanical robustness tests, aiming at implementation in an electric vehicle for the Formula Student competition.

Author Contributions: Conceptualization, J.G.O.P. and J.P.D.M.; methodology, J.G.O.P., J.P.D.M. and L.A.M.B.; software, J.P.D.M.; validation, J.P.D.M., J.G.O.P. and L.A.M.B.; formal analysis, all authors; investigation, J.G.O.P., J.P.D.M. and L.A.M.B.; resources, J.G.O.P.; data curation, all authors; writing—original draft preparation, J.G.O.P.; writing—review and editing, all authors; visualization, all authors; supervision, J.G.O.P. All authors have read and agreed to the published version of the manuscript.

Funding: This work has been supported by FCT—Fundação para a Ciência e Tecnologia within the R&D Unit Project Scope UID/00319/Centro ALGORITMI (ALGORITMI/UM).

Data Availability Statement: The original contributions presented in the study are included in the article, further inquiries can be directed to the corresponding authors.

Conflicts of Interest: The authors declare no conflicts of interest.

Abbreviations

The following abbreviations are used in this manuscript:

BMS	battery management system
CAN	controller area network
dc	direct current
DoD	depth of discharge
EV	electric vehicle
MOSFET	metal oxide semiconductor field effect transistor
PWM	pulse width modulation
SoC	state of charge
SoH	state of health

References

1. Tintelecan, A.; Dobra, A.C.; Marţiş, C. LCA Indicators in Electric Vehicles Environmental Impact Assessment. In Proceedings of the 2019 Electric Vehicles International Conference (EV), Bucharest, Romania, 3–4 October 2019; pp. 1–5. [CrossRef]
2. Maher, K. Study of 18650 $LiFePO_4$ Lithium-Ion Battery Cells for Stationary and EV Applications: Performance, Reliability, and Safety. In Proceedings of the 2023 6th International Conference on Renewable Energy and Power Engineering (REPE), Beijing, China, 15–17 September 2023; pp. 156–162. [CrossRef]
3. Fares, A.; Klumpner, C.; Sumner, M. Investigating the benefits and limitations of cascaded converter topologies used in modular battery systems. In Proceedings of the 2017 IEEE 26th International Symposium on Industrial Electronics (ISIE), Edinburgh, UK, 19–21 June 2017; pp. 2123–2130. [CrossRef]
4. Kumar, S.; Jawaharram, V.R.; R, B.; Chandrakala, K.R.M.V.; Kumar, S.S.; Nadar, K.P. Scalable Modular Battery Management System Using Model-Based Approach for EV Applications. In Proceedings of the 2024 Second International Conference on Emerging Trends in Information Technology and Engineering (ICETITE), Vellore, India, 22–23 February 2024; pp. 1–5. [CrossRef]
5. Vijaychandra, J.; Knypiński, Ł. A Comprehensive Review on Challenges and Possible Solutions of Battery Management Systems in Electric Vehicles. In Proceedings of the 2024 Progress in Applied Electrical Engineering (PAEE), Koscielisko, Poland, 24–28 June 2024; pp. 1–6. [CrossRef]
6. Miranda, J.P.D.; Barros, L.A.M.; Pinto, J.G. A Review on Power Electronic Converters for Modular BMS with Active Balancing. *Energies* **2023**, *16*, 3255. [CrossRef]
7. Chauhan, S.R.; Kumar, K.; Nadarajan, S.; Vaiyapuri, V.; Halick, M.; Sathik, M. Effect of Unbalanced Cells in Lithium-ion Battery Pack Performance and SOC Estimation. In Proceedings of the 2024 12th International Conference on Internet of Everything, Microwave, Embedded, Communication and Networks (IEMECON), Jaipur, India, 24–26 October 2024; pp. 1–6. [CrossRef]
8. Ashraf, A.; Ali, B.; Alsunjury, M.S.A.; Goren, H.; Kilicoglu, H.; Hardan, F.; Tricoli, P. Review of Cell-Balancing Schemes for Electric Vehicle Battery Management Systems. *Energies* **2024**, *17*, 1271. [CrossRef]
9. Zhou, G.; Zhang, X.; Gao, K.; Tian, Q.; Xu, S. Two-Mode Active Balancing Circuit Based on Switched-Capacitor and Three-Resonant-State LC Units for Series-Connected Cell Strings. *IEEE Trans. Ind. Electron.* **2022**, *69*, 4845–4858. [CrossRef]

10. Panchal, A.; Bhatt, K.; Gitaye, S.; Bhand, M.; Sheikh, A. Design and Simulation of an Inductor based Active Cell Balancing Circuit for Lithium-ion Batteries. In Proceedings of the 2022 4th Global Power, Energy and Communication Conference (GPECOM), Nevsehir, Turkey, 14–17 June 2022; pp. 89–94. [CrossRef]
11. Manjunath, K.; R, K. Active Cell Balancing Circuit using Switched inductor Buck-Boost Converter for Li-ion Battery Strings with Maximum Efficiency Operation. In Proceedings of the 2023 IEEE International Conference on Power Electronics, Smart Grid, and Renewable Energy (PESGRE), Trivandrum, India, 17–20 December 2023; pp. 1–6. [CrossRef]
12. Kipke, V.; Dost, P.; Sourkounis, C.; Widera, B. Novel Direct Active Cell-to-Cell Balancing Approach for Energy Storage Systems based on a Flying Inductor Circuit Topology. In Proceedings of the IECON 2024—50th Annual Conference of the IEEE Industrial Electronics Society, Chicago, IL, USA, 3–6 November 2024; pp. 1–8. [CrossRef]
13. Lee, S.; Kim, M.; Baek, J.W.; Kang, D.-W.; Jung, J. Enhanced Switching Pattern to Improve Cell Balancing Performance in Active Cell Balancing Circuit Using Multi-Winding Transformer. *IEEE Access* **2020**, *8*, 149544–149554. [CrossRef]
14. Srinivasan, M.P.; Parimi, A.M. Design of an Isolated Bidirectional Active Cell Balancing circuit for Lithium ion batteries. In Proceedings of the 2022 IEEE 7th International Conference for Convergence in Technology (I2CT), Mumbai, India, 7–9 April 2022; pp. 1–6. [CrossRef]
15. Conway, T. An Isolated Active Balancing and Monitoring System for Lithium Ion Battery Stacks Utilizing a Single Transformer Per Cell. *IEEE Trans. Power Electron.* **2021**, *36*, 3727–3734. [CrossRef]
16. Kwon, J.-H.; Choi, S.-C.; Zhou, G.; Park, S.-M.; Park, S.-J.; Wang, Y. A Cell-to-Cell Voltage Balancing Strategy with Bidirectional Flyback Converter. In Proceedings of the 2024 IEEE 10th International Power Electronics and Motion Control Conference (IPEMC2024-ECCE Asia), Chengdu, China, 17–20 May 2024; pp. 4088–4093. [CrossRef]
17. Pinto, J.G.O.; Silva, M.F.; Barros, L.A.M.; Afonso, J.A. New BMS Topology with Active Cell Balancing Between Electric Vehicles' Traction and Auxiliary Batteries. *Batteries* **2025**, *11*, 175. [CrossRef]
18. Ye, Y.; Lin, J.; Li, Z.; Wang, X. Double-Tiered Cell Balancing System with Switched-Capacitor and Switched-Inductor. *IEEE Access* **2019**, *7*, 183356–183364. [CrossRef]
19. Jain, A.; Jain, S.K. Multi-Layer Cell Balancing Using Switched Inductor and Switched Capacitor Topology. In Proceedings of the 2021 28th International Workshop on Electric Drives: Improving Reliability of Electric Drives (IWED), Moscow, Russia, 27–29 January 2021; pp. 1–6. [CrossRef]
20. Al-Smadi, M.K.; Qahouq, J.A.A. Generalized Analysis and Evaluation of Switched Inductor PWM-based Lithium-ion Battery Cell Balancing. In Proceedings of the 2022 IEEE Energy Conversion Congress and Exposition (ECCE), Detroit, MI, USA, 9–13 October 2022; pp. 1–5. [CrossRef]
21. Moghaddam, A.F.; Van Den Bossche, A. An Active Cell Equalization Technique for Lithium Ion Batteries Based on Inductor Balancing. In Proceedings of the 2018 9th International Conference on Mechanical and Aerospace Engineering (ICMAE), Budapest, Hungary, 10–13 July 2018; pp. 274–278. [CrossRef]
22. Yang, W.; Han, J.; Liu, X.; Yang, S. An active cell-to-cell balancing circuit with planar coupled inductors for series connected batteries. In Proceedings of the 2019 IEEE 4th International Future Energy Electronics Conference (IFEEC), Singapore, 25–28 November 2019; pp. 1–6. [CrossRef]
23. Ye, Y.; Jiang, J.; Zhao, E.; Li, P.; Li, Z.; Hui, X. An Improved Balancing Strategy for Inductor-Based Balancing Circuit. In Proceedings of the 2023 IEEE International Conference on Power Science and Technology (ICPST), Kunming, China, 5–7 May 2023; pp. 387–391. [CrossRef]
24. Frivaldsky, M.; Simcak, M.; Andriukaitis, D.; Navikas, D. Cycling Operation of a $LiFePO_4$ Battery and Investigation into the Influence on Equivalent Electrical Circuit Elements. *Batteries* **2025**, *11*, 211. [CrossRef]

batteries

MDPI

Article

In Situ Estimation of Li-Ion Battery State of Health Using On-Board Electrical Measurements for Electromobility Applications

Jorge E. García Bustos [1,*], Benjamín Brito Schiele [1], Leonardo Baldo [2], Bruno Masserano [1], Francisco Jaramillo-Montoya [1], Diego Troncoso-Kurtovic [3], Marcos E. Orchard [1], Aramis Perez [4] and Jorge F. Silva [1]

1 Department of Electrical Engineering, Faculty of Physical and Mathematical Sciences, University of Chile, Av. Tupper 2007, Santiago 8370451, Chile; benjamin.brito@ug.uchile.cl (B.B.S.); bruno.masserano@ug.uchile.cl (B.M.); francisco.jaramillo@uchile.cl (F.J.-M.); morchard@ing.uchile.cl (M.E.O.); josilva@ing.uchile.cl (J.F.S.)
2 Department of Mechanical and Aerospace Engineering, Politecnico di Torino, Corso Duca degli Abruzzi 24, 10129 Torino, Italy; leonardo.baldo@polito.it
3 Center for Sustainable Acceleration of Electromobility (CASE), Faculty of Physical and Mathematical Sciences, University of Chile, Santiago 8370451, Chile; diego.troncoso.k@ug.uchile.cl
4 The School of Electrical Engineering, University of Costa Rica, San José 11501-2060, Costa Rica; aramis.perez@ucr.ac.cr
* Correspondence: jorgegarcia@ug.uchile.cl

Abstract

The well-balanced combination of high energy density and competitive cycle performance has established lithium-ion batteries as the technology of choice for Electric Vehicles (EVs) energy storage. Nevertheless, battery degradation continues to pose challenges to EV range, safety, and long-term reliability, making accurate estimation of their State of Health (SoH) crucial for efficient battery management, safety, and improved longevity. This paper addresses a compelling research question surrounding the possibility of developing a real-time, non-invasive, and efficient methodology for estimating lithium-ion battery SoH without battery removal, relying solely on voltage and current data. Our approach integrates the fitting abilities of Maximum Likelihood Estimation (MLE) with the dynamic uncertainty propagation of Bayesian Filtering to provide accurate and robust online SoH estimation. By reconstructing the open-circuit voltage curve from real-time data, the MLE estimates battery capacity during discharge cycles, while Bayesian Filtering refines these estimates, accounting for uncertainties and variations. The methodology is validated using an available dataset from Stanford University, demonstrating its effectiveness in tracking battery degradation under driving profiles. The results indicate that the approach can reliably estimate battery SoH with mean absolute errors below 1%, confirming its suitability for scalable EV applications.

Keywords: electric vehicles; lithium-ion batteries; battery degradation; maximum likelihood estimation; particle filtering

1. Introduction and Motivation

Driven by the urgent need to decarbonize transportation and minimize its environmental impact, Electric Vehicles (EVs) are slowly but steadily reshaping the mobility paradigm, offering an accessible and sustainable alternative to conventional internal combustion systems. One of the key enablers of the shift towards electromobility is the substantial progress in battery technology, whose improved autonomy has significantly contributed to EVs competitiveness. In this sense, Lithium-Ion Batteries (LIBs) have established themselves as the primary energy source for EVs due to their high energy density and relatively long cycle life compared to other battery technologies [1,2]. Yet, despite these advantages, LIB degradation remains a critical concern, especially when accelerated by challenging operating conditions such as extremely high discharge/charge rates, extreme temperatures, and deep depth of discharge (DoD) [3]. In EVs, this performance degradation directly impacts vehicle autonomy, intensifies range anxiety, and increases the probability of thermal runaway events [4,5]. Thus, it is clear that real-time monitoring of battery health under operational conditions is of paramount importance to both the individual user and the Original Equipment Manufacturer, which can benefit from improved situational awareness and performance management [6].

In the literature, battery condition is commonly associated with the notion of State of Health (SoH), a metric that relates the current battery capacity to its initial nominal capacity [7]. Although intuitively straightforward, SoH cannot be measured directly and requires estimation techniques, which are often impractical, time-intensive, or incompatible with EV battery systems, as they would require the removal of the battery pack from the vehicle [8]. Alternatively, a conventional solution to obtain SoH with the battery still loaded in the vehicle involves performing a full discharge from a fully charged battery, followed by a rest period and a subsequent full recharge [9]. Based on the Coulomb counting strategy, this methodology offers precise SoH measurements; however, it requires the vehicle to be stationary during the long testing process. Alternative strategies, such as Electrochemical Impedance Spectroscopy (EIS), have been proposed to mitigate the time-consuming limitations of Coulomb counting [10,11]; however, they often involve battery disassembly or off-board testing, restricting their practicality in operational settings [12]. At the same time, tests that require access to cell-level data are unfeasible as such data are rarely available through EV's Battery Management Systems (BMSs) [13]. In this sense the main category of SoH estimation methods based on on-board electrical measurements is represented by Differential Analysis Methods. These techniques, often divided into direct and indirect measurements, require laboratory setups and specialized equipment and, as a result, are not feasible for online applications [14], which is why they were excluded from the present review. In summary, the constant assessment of battery SoH via the aforementioned strategies conflicts with intrinsic conditions of the battery installed on EVs, as the existing estimation methodologies require disconnecting the battery or time-consuming procedures to accurately assess its health status, impeding estimations "on the go".

Over the years, some methodologies have been proposed for online SoH estimation in real-world applications based on phenomenological principles, expert knowledge, and/or operational data of the cell/battery when available [15,16]. Similarly, the literature includes proposals addressing operational uncertainty and computational efficiency. These are particularly relevant to review given the wide range of variations that may arise in electromobility applications and the need to consider on-board implementations for SoH estimation.

In the case of data-driven approaches that leverage operational data, illustrative examples are provided by Li et al. [17] and Ghosh et al. [18]. On the one hand, [17] uses empirical mode decomposition and Pearson Correlation coefficients to extract Health Indicators (HIs) from measurements obtained during the charging process, subsequently processed through long short-term memory (LSTM) and BiLSTM networks. On the other hand, Ref. [18] employ an LSTM network for voltage curve reconstruction, then feeds these voltage predictions alongside measured current in a quantum-fuzzy neural network (eQFNN) framework. Both studies achieve high accuracy but encounter some limitations:

Ref. [17] requires future information not commonly available in real-world applications, while the approach of [18] may suffer from excessive computational complexity that challenges onboard implementation. Other interesting examples are found in Feng et al. [19], which developed a Support Vector Machine (SVM) method, which utilizes partial constant-current charging segments to compare measured voltage profiles with a pre-learned SVM library. Moreover, Etxandi-Santolaya et al. [20] estimates SoH and battery functionality using Support Vector Regression and neural networks trained on HIs extracted from partial charging segments. Despite the scalability and performance under realistic conditions, in both cases, the methods are defined at a cell level and do not consider the variability, filtering, cell balancing, or constraints introduced by BMS hardware or firmware in the data acquisition process.

For hybrid approaches that combine data-driven and physics-based methods, Zhang et al. [21] provide a representative example. Their work seeks to integrate the flexibility of data-driven models with physical constraints through a two-stage framework. First, a Gated Recurrent Unit (GRU) is used to give an initial value of SoH from historical data. Subsequently, a Particle Filter (PF) adjusts the SoH and provides State of Charge (SoC) estimations. This method can adapt the SoH in real-time, but it depends only on the GRU for the initial estimation: errors in initialization propagate through the system, requiring considerable time to be fixed by the PF.

While previous methodologies have been deterministic, probabilistic approaches have also been proposed to explicitly account for uncertainty. Despite their diversity, these approaches tend to converge toward comparable underlying theoretical frameworks. For instance, the study of Dong et al. [22] proposes a Hybrid Kernel Relevance Vector Machine (H-RVM) for SoH and Remaining Useful Life (RUL) estimation. This approach can achieve good results at a practical level, but it depends on keeping consistent charging protocols, which limits the applicability, given the different charging strategies in different charging stations. In turn, Xiang et al. [23] incorporate Gaussian Process Regression (GPR) to quantify uncertainty, focusing on discharge segments with pseudo-labeling in a semi-supervised bidirectional GRU framework, while Wang et al. [24] employ GPR directly for nonlinear process modeling of the relationship between measurable battery features, proposing an improved Firefly Algorithm (IFA) into the training process to improve global search capability. The methodology presented in [23] can be applied in an online setting; however, it may introduce considerable errors, as the GRU loses access to future data points and potential mislabeling can propagate inaccuracies in subsequent estimations. On the other hand, in [24], the method can encounter high computational costs and a strong reliance on offline training data. In conclusion, probabilistic methods excel in uncertainty quantification and nonlinear modeling but require high-quality data and consistent requirements, which can be challenging in practice.

Another dimension to be reviewed is related to the data flow efficiency question for SoH estimation. One effort is presented in He et al. [25], where the SoH estimation is based on the variation coefficient extracted from partial charging curves and a GPR. The resulting framework keeps predictions coherent across different training sizes and enables online parameter updates that shorten commissioning time. A complementary path is explored in the study by Buchanan and Crawford [26], where a convolutional encoder learns directly from raw charge traces and a sparse GPR head outputs probabilistic SoH trajectories with explicit credibility bands. In Dong et al. [27], the concept of probabilistic co-estimation is further advanced. The authors calibrate the parameters of a fractional-order equivalent circuit model using Bayesian optimization and subsequently integrate a Gaussian-sum particle filter with recursive total least squares to jointly update the SoH and SoC. This approach eliminates the need for laboratory impedance tests and aligns with firmware-level identification routines. In this regard, it is worth noting that, consistent with broader trends in Prognostics and Health Management (PHM), hybrid methodologies that combine data-driven and model-based approaches are emerging as a promising direction for SoH estimation, as emphasized by Zhang et al. [21] and Deng et al. [28]. Lastly, the paper by Zhu et al. [29] brings Bayesian calibration into the second life scenario using a physics-informed neural network that is pre-trained on porous electrode simulations and refined with Monte Carlo dropout, achieving robust estimates across heterogeneous aging paths and pointing to lighter uncertainty-aware estimators suitable for embedded controllers.

Inspired by these findings, we propose a hybrid estimation method that integrates model-based approaches, parametric estimation, and Bayesian inference to estimate the current SoH during EV operation, using non-invasive data from onboard systems such as voltage and current trends from the Controller Area Network (CAN) bus. The proposed solution uses maximum likelihood estimation (MLE) to extract capacity estimates from readily available online operational data, utilizing these measurements without requiring invasive diagnostic techniques. To further strengthen the robustness and resilience of the estimation process, a Bayesian Filter (BF) loop is added to the framework. In particular, the PF is employed as an implementation of the BF problem. In fact, MLE alone can be sensitive to noise, short discharge segments, and sensor inaccuracies. To address these challenges, estimation is further enhanced by the application of a PF, which smooths estimates, compensates for uncertainties, and mitigates errors caused by sensor anomalies or missing data. The comprehensive combined MLE-PF approach provides a robust and adaptive estimator that improves accuracy by integrating real-time measurements with a degradation model, ensuring reliable SoH estimation even in the presence of quantized or noisy operational data. In this work, we assume that the EV is continuously monitored by an onboard computer or by the BMS. The only data requirement is access to operational signals, such as voltage and current trends available on the CAN bus, enabling seamless online estimation of both SoH and SoC.

The main contributions of this article are:

1. A novel MLE-PF battery capacity estimation procedure is developed to obtain SoH estimates from an arbitrary current and voltage discharge profile.
2. A PF algorithm is implemented to overcome model and estimation uncertainties and potential outliers during the degradation process.
3. The addition of the PF enables the simultaneous online estimation of SoH and SoC without the need for time-intensive tests, which require battery disconnection.
4. The proposed method is designed for use on a generic EV since current and voltage measurements are easy to obtain from a BMS or an OBD2 device.
5. The proposed approach can be scaled to different types, modules, and lithium ion batteries since the parameters of the battery model can be adjusted with few operational data.

The remainder of this paper is organized as follows. Section 2 provides a comprehensive theoretical background, discussing key concepts such as SoC, SoH, and the open-circuit voltage (OCV) model. Section 3 details the proposed methodology, including a description of the MLE and its integration with the PF algorithm. Section 4 introduces a case study to demonstrate the applicability of the proposed approach, utilizing real-world data from LIB cells. Section 5 presents and analyzes the results, highlighting the performance of the MLE-PF strategy in accurately estimating battery health under various operating conditions. Finally, Section 6 summarizes the findings, discusses implications for future research, and concludes the paper by highlighting the potential of the proposed method for real-time applications in EVs.

2. Theoretical Background

In this section, we present the fundamental concepts and methodologies that form the basis for the proposed approach in this paper. First, we provide essential definitions and concepts related to battery condition monitoring, including the SoC, SoH, and the OCV model. Then, we introduce key statistical techniques such as the MLE, BF/PF, which are instrumental in refining the accuracy of SoH and SoC estimates.

2.1. Definitions and Equivalent Circuit Battery Model

The SoC and SoH are crucial indicators for evaluating battery operation, and they describe the state of the battery from two fundamentally different points of view. The SoC describes an intercharge performance, while the SoH represents an intracharge performance. That is, the SoC describes the charge available with respect to the maximum charge available at that specific moment, while the SoH represents the trend of the maximum capacity of the battery with respect to the initial capacity of the new battery. As such, these concepts also refer to two completely different timescales, identified with the indices t and k. t is the time step index used in the SoC calculation, while k is the work cycle index used in the SoH calculation to indicate operational charge-discharge cycles, defined as a process during which the battery delivers an amount of energy equivalent to 100% of its current capacity, given its SoH [7].

1. The SoC represents the current capacity of the battery relative to its total current capacity, typically expressed as a percentage [30]. It indicates how much charge remains in the battery and its evolution in time can be mathematically formulated as:

$$SoC_t = SoC_{t-1} - \frac{I_{t-1} \cdot \Delta t_t}{C_k} + w_t\,, \tag{1}$$

where SoC_{t-1} and I_{t-1} are the SoC and the current at the previous time step, respectively, Δt_t is the sampling time interval, C_k is the battery capacity at cycle k (i.e., the maximum charge the cell can store at that cycle), and w_t represents the process noise accounting for uncertainties such as current sensor inaccuracies, unmodeled dynamics, and environmental variations. By incorporating noise, the model becomes more robust and can better capture the real-world behavior of batteries, enhancing the credibility and reliability of the SoC estimate. The SoC trend can vary over time depending on the discharge rate, which in practice is associated with different usage profiles [31].

To connect the SoC concept with practical battery applications, it is essential to consider an equivalent circuit model that simulates the battery behavior under various operational scenarios. In this work, we use a Thévenin-equivalent model, which includes a controlled voltage source that represents the open-circuit voltage V_{oc}, integrated with a resistance to simulate real-world conditions [32].

The open-circuit voltage V_{oc} is a nonlinear function of the SoC and is can be represented through a three-stage structure that reflects the dominant electrochemical mechanisms across different SoC ranges. Studies such as [33] have shown that, at high SoC values, a combination of partial redox reactions and increased charge accumulation near electrode saturation produces a pronounced curvature in the OCV–SoC relationship, effectively described by a shifted exponential function. In the mid-SoC region, the main intercalation and phase-transition reactions of the active materials proceed smoothly, yielding a nearly affine voltage response that can be captured by a linear term. Conversely, at low SoC values, redox activity is minimal and the voltage exhibits a sharp decline driven by surface charge-accumulation effects, which is well modeled by logarithmic or exponential behavior.

Beyond these chemical considerations, it is also essential to account for the aging-induced voltage drift observed in the OCV curve as the battery degrades. As reported in [34], periodic OCV measurements exhibit a progressive leftward shift over the cell's lifetime, reflecting both capacity loss and changes in the electrode equilibrium potentials. This systematic displacement indicates that a fixed OCV curve cannot accurately represent the evolving equilibrium voltage associated with different SoH levels, thereby motivating the use of a parametric formulation whose coefficients can be adapted to aging.
Building on this structure and considering the proposed Thévenin-equivalent circuit, the five-parameter OCV model introduced in [35] is adopted in this work and expressed in Equation (2) where each term in corresponds to a distinct operational zone of the OCV curve. The offline procedure proposed in [36] allows the identification of model parameters using a single controlled-discharge experiment, whereas [37] extends this approach to data collected from an EV operating under real driving conditions. These formulations enable the OCV curve to adapt to different SoH levels, as illustrated in Figure 1.

$$v_{oc}(SoC_t) = v_L + (v_0 - v_L) \cdot e^{\gamma(SoC_t - 1)} + \alpha \cdot v_L(SoC_t - 1) + (1 - \alpha) v_L (e^{-\beta} - e^{-b\sqrt{SoC_t}}). \quad (2)$$

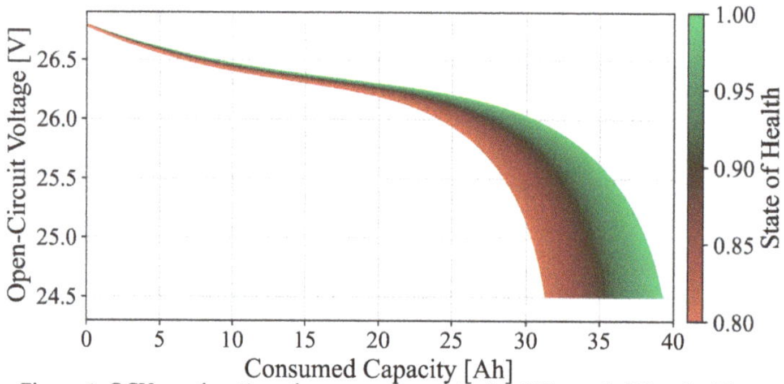

Figure 1. OCV as a function of consumed capacity for different SoH levels. The transition from green (higher SoH) to red (lower SoH) illustrates how decreasing SoH alters the OCV curve and reduces the usable capacity.

2. On the other hand, the SoH refers to the battery's ability to deliver its total current capacity compared to its original (or nominal) capacity when new [30]. In the battery literature, SoH can be characterized through several degradation indicators (e.g., capacity loss, increase in internal impedance, power fade, or self-discharge rate) which contribute to describing the battery health state [3]. Nevertheless, many studies point out that the most widely adopted formulation in both research and practice remains the capacity-based definition [38,39]. That is why, in the context of this study, we adopt the most common and widely accepted formulation, and we define SoH solely on a capacity basis, as reported in Equation (3).

$$SoH_k = \frac{C_k}{C_0}, \quad (3)$$

where SoH_k is the State of Health at cycle k, C_k is the battery capacity at cycle k, and C_0 is the nominal (initial) capacity when the battery is new. The capacity C_k decreases with each cycle due to degradation, affecting the duration and energy delivered in subsequent cycles. A common way to represent degradation over the battery operational cycles is by describing the maximum energy storage capability that the battery can deliver over time. One way to model this degradation is through the equation:

$$C_k = C_{k-1} \cdot \eta_{k-1} + v_k, \tag{4}$$

where C_k is the capacity at the current cycle, C_{k-1} and η_{k-1} are the capacity and the efficiency factor in the previous cycle, respectively. and v_k represents the process noise capturing uncertainties such as degradation rate variability, measurement errors, and unmodeled aging effects. Similarly to w_t, the term v_k is introduced to model the inherent uncertainties and stochastic nature of battery operation to make SoH estimation more robust.

The efficiency factor η serves as an aggregate representation of the Loss of Lithium Inventory (LLI) induced by each duty cycle experienced by the battery [40]. As highlighted in [41], LLI is one of the dominant degradation modes in LIBs, arising primarily from SEI growth, lithium plating, and crack-assisted side reactions, all of which irreversibly trap cyclable lithium and reduce the amount of charge that can be reversibly stored. These mechanisms occur even under moderate operating conditions and accumulate over time, leading to measurable reductions in accessible capacity. By expressing the degradation of a given cycle as a multiplicative efficiency applied to the previous available capacity, the model effectively captures the fraction of lithium lost to these irreversible pathways, allowing the model to track the progressive depletion of cyclable lithium as the battery ages.

Since the SoC is intrinsically linked to the SoH via the current maximum capacity, the SoH also influences the behavior of the V_{oc} curve [42] as an external parameter. Figure 1 illustrates this dependency by showing how the OCV curve shifts with different SoH values.

In fact, a reduced battery capacity implies that the SoC will drop more quickly for the same amount of extracted charge, causing V_{oc} to decrease significantly faster. Starting from Equation (2), this relationship is expressed mathematically in Equation (5), where it is clear that V_{oc} is a function of the SoC. More specifically, C_k represents the battery capacity at a given cycle k and Q_{ext} the extracted charge. Note that a lower C_k increases the impact of Q_{ext} on V_{oc}.

$$V_{oc} = v_{oc}(SoC) = v_{oc}\left(\frac{C_k - Q_{\text{ext}}}{C_k}\right) = v_{oc}\left(1 - \frac{Q_{\text{ext}}}{C_k}\right) \tag{5}$$

2.2. Estimation

In this section, we provide an overview of the key estimation techniques used in this paper. Estimation plays a crucial role in determining the state of a dynamic system, such as the SoC and SoH of a battery. The methods discussed here, namely MLE and BF, with a focus on PF, serve as powerful tools for accurately inferring these states based on available data. By combining these techniques, we aim to develop a more robust and precise estimation framework, adaptable to different scenarios and model structures.

2.2.1. Maximum Likelihood Estimation

MLE is a technique widely used in statistics and engineering applications to estimate the parameters of a model that maximizes the probability of occurrence of a given event [43–45]. The basic principle of MLE is to define a likelihood function, $L(\theta; x)$, which represents the probability of observing the event $\mathbf{x}$ given a set of parameters θ. The goal of MLE is to find the values of θ that maximize the likelihood function [46], mathematically expressed as:

$$\hat{\theta} = \underset{\theta}{\operatorname{argmax}}\ L(\theta; \mathbf{x}). \tag{6}$$

It is customary to work with the log-likelihood function due to its level of mathematical convenience, since converting the product of probabilities into a sum simplifies the calculations. This can be represented by:

$$\log L(\theta; \mathbf{x}) = \sum_{i=1}^{n} \log f(x_i; \theta), \tag{7}$$

where $f(x_i; \theta)$ corresponds to the probability density function of the data x_i given the parameters θ.

2.2.2. Bayesian Filtering

BF is a statistical technique employed to estimate and predict the state of a dynamic system based on a given model and noisy measurements. It relies on Bayes' theorem to update the probability distribution of the system state as new data becomes available [47]. In particular, an iterative algorithm predicts the next state using the system model, the current state, an exogenous variable, and process noise (Predictive step). This can be expressed in general terms as follows:

$$x_k = f_k(x_{k-1}, u_{k-1}, \omega_{k-1}), \tag{8}$$

where $f(\cdot)$ is a function that depends on the state vector x, the input $u \in \mathbb{R}^{n_u}$, and the i.i.d process noise vector $\omega \in \mathbb{R}^{n_\omega}$. In the second step (Update), this prediction is used to compute what should be expected according to the measurement equation:

$$z_k = h_k(x_k, v_k), \tag{9}$$

where $h(\cdot)$ is a function, $z \in \mathbb{R}^{n_z}$ is the measurement vector, and $\nu \in \mathbb{R}^{n_\nu}$ is the noise vector [47].

Finally, the likelihood of the observed measurement is computed and used to update the state and calculate the uncertainty, assuming the availability of the initial condition $p(\mathbf{x}_0 \mid \mathbf{z}_0) = p(\mathbf{x}_0)$. This process is iterated until an end criterion is met.

Particle Filter

PF are algorithms that offer a solution to the BF problem by using particles, which are defined as a set of N_p random samples with their respective weights (see Equation (10)), and Monte Carlo simulations, which are employed to approximate the posterior distribution of the desired state, since the true distribution is often intractable [48].

$$\left\{\mathbf{x}_k^{(i)}, w_k^{(i)}\right\}_{i=1}^{N_p}, \quad \sum_{i=1}^{N_p} w_k^{(i)} = 1 \tag{10}$$

PF samples the N_p particles at every step from an alternative PDF $q(\cdot)$ named importance density [49]. Therefore, the posterior distribution can be approximated by [50,51]:

$$p(\mathbf{x}_k|\mathbf{z}_{1:k}) \approx \sum_{i=1}^{N_p} w_k^{(i)} \delta(\mathbf{x}_k - \mathbf{x}_k^{(i)}), \tag{11}$$

where the weights of each particle are updated as follows:

$$w_k^{(i)} = w_{k-1}^{(i)} \frac{p(\mathbf{z}_k|\mathbf{x}_k^{(i)}) p(\mathbf{x}_k^{(i)}|\mathbf{x}_{k-1}^{(i)})}{q(\mathbf{x}_k^{(i)}|\mathbf{x}_{0:k-1}^{(i)}, \mathbf{z}_{1:k})} \tag{12}$$

It is crucial to highlight that Equation (11) converges to an equality as $N_p \to \infty$. Furthermore, the design and performance of the PFs are significantly influenced by the choice of the importance density $q(\cdot)$ [52]. In this study, the Sequential Importance Sampling (SIS) PF was implemented, which assumes the following equality:

$$q(\mathbf{x}_k^{(i)}|\mathbf{x}_{k-1}^{(i)}, \mathbf{z}_k) = p(\mathbf{x}_k|\mathbf{x}_{k-1}), \tag{13}$$

indicating that the $q(\cdot)$ probability density function (PDF) is equal to the prior PDF. For further details on SIS, the interested reader can find an in-depth description of the methodology in [53–55].

The SIS approach was adopted for its conceptual simplicity and computational efficiency. By setting the prior transition model as the importance density, SIS avoids the need to approximate complex proposal distributions, which are often intractable in nonlinear or high-dimensional systems. This assumption reduces computational cost while providing a consistent approximation of the posterior distribution, although it may suffer from weight degeneracy over long sequences [50,52]. To mitigate the degeneracy issue, a resampling step is introduced when the Effective Sample Size falls below a threshold [50,56]. This step removes particles with negligible weights and replicates those with high importance, thereby improving numerical stability and filter robustness, albeit with a slight increase in sampling variance [57].

3. MLE-PF Framework: A Simple and Reliable Estimation Method for Battery Capacity Using Operational Data

3.1. Theoretical Rationale for Battery Capacity Estimation

As shown in Equation (3), the SoH is directly determined by the battery capacity. Therefore, knowing the initial capacity C_0 and the capacity at a given cycle C_k, the SoH at cycle k can easily be derived.

Considering the relationship between V_{oc} and the battery capacity presented in Equation (5), an estimator can collect a set of V_{oc} and extract charge measurements to infer the capacity value that best fits the observations, thus providing the needed value to calculate the SoH. The optimization problem can be formalized in Equation (14), which represents the core of the framework proposed in this paper.

$$\hat{C} = \arg\min_{C} \sum_{i=1}^{N} \left[V_{\text{oc}}^{(i)} - v_{\text{oc}}\left(1 - \frac{Q_{\text{ext}}^{(i)}}{C}\right)\right]^2, \tag{14}$$

where $V_{oc}^{(i)}$ and $Q_{ext}^{(i)}$ represent the i-th V_{oc} and extracted charge measurements, whereas v_{oc} refers to the parametric Equation (2).

The direct application of this inference strategy poses several challenges, primarily due to the fundamental operating principles of a battery when connected within an electrical circuit. These challenges must be addressed to ensure reliable implementation. The primary problem lies in the need for accurate open-circuit voltage V_{oc} measurements. In fact, to acquire accurate readings that are not affected by internal impedance, extended rest periods without current flow are required. Furthermore, accurately quantifying the extracted charge Q_{ext} requires the discharge process to start from a fully charged state. However, in real-world scenarios, this condition is rarely met, as drivers do not always start their trips with a fully charged battery.

Overcoming these challenges requires an integrated framework capable of reconstructing voltage values that closely approximate the true open-circuit voltage V_{oc} using real-world measurements of battery voltage and current during EV operation.

3.2. Framework Overview

Based on the information obtained in the previous section, a Maximum Likelihood Estimation-PF (MLE-PF) approach has been implemented to estimate the battery capacity using Equation (14), as illustrated in Figure 2. In particular, vehicle data are initially passed through a Feature Extraction Module (FEM) to generate a set of observations, which the MLE module then uses to estimate the battery capacity. On the other hand, a parallel model-based branch (highlighted in red) is integrated starting from vehicle data. The two branches converge into the PF, which fuses observations and model predictions to output a filtered battery capacity estimate, from which the SoH is subsequently computed. The inclusion of the PF aims to obtain a more reliable and robust SoH estimator since it helps to overcome off-trend estimates, which may be the result of having to deal with low-quality or noisy operational data.

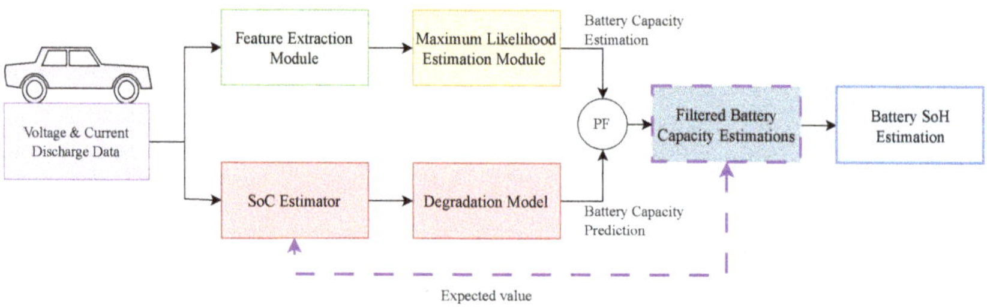

Figure 2. Vehicle data are processed by the FEM to generate relevant observations, which are then supplied to the MLE Module for battery capacity estimation. A parallel model-based branch (in red) supplies a degradation model to the PF, which fuses observations and predictions to produce a filtered capacity estimate used to determine the SoH.

Figure 3 shows a detailed view of the MLE-PF framework, focusing on the FEM and MLE module. The FEM logs voltage and charge measurements from the natural discharge process during standard EV operation. Measurements that meet a set of predefined criteria are retained and processed for further analysis. Once a meaningful set of observations is harvested, the acquired dataset is passed to the MLE module, which, by solving an estimation process, estimates the actual current battery capacity and therefore the SoH. This process can be run online while driving.

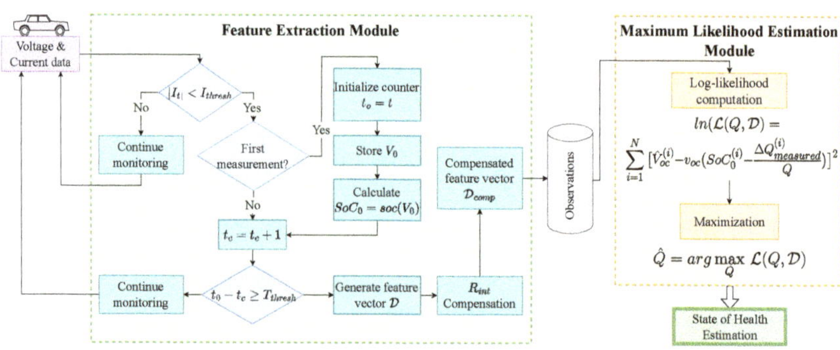

Figure 3. Flowchart illustrating the Feature Extraction and Maximum Likelihood Estimation modules employed to estimate battery capacity and State-of-Health.

3.2.1. Feature Extraction Module

The FEM operates by identifying and collecting informative voltage and charge measurements during battery discharge cycles that occur under typical electric vehicle operating conditions. The primary objective of the FEM is to identify voltage readings that closely approximate the open-circuit voltage V_{oc}. These selected measurements form the basis for estimating the battery current capacity. In fact, the direct correlation between open-circuit voltage and battery capacity demonstrated in Equation (14) gives a clear strategy to calculate the current capacity, starting from the V_{oc}.

However, since the battery is always connected to the circuit and under load, a perfect reconstruction of the true V_{oc} is inherently unachievable; therefore, the resulting voltage can only approximate the true open-circuit condition. For this reason, the reconstructed value is referred to as a pseudo-open-circuit voltage or V_{ocp}. While in operation, the measured voltage can be described by the relationship:

$$V_{ocp} = V_t + I_t \cdot R_{int_t}, \tag{15}$$

where I_t, V_t, and SoC_t correspond to the current, voltage, and SoC at the t-th discharge instant, respectively. Finally, R_{int_t} represents the internal resistance of the battery. To obtain voltage measurements that closely resemble true open-circuit conditions, the FEM must reduce the influence of the voltage drop caused by the $I_t \cdot R_{int_t}$ term. When this objective is achieved, the pseudo-open-circuit voltage approximates the actual open-circuit voltage:

$$V_{ocp} \approx V_{oc} = v_{oc}(SoC_t) \tag{16}$$

It is important to note that Equation (16) maintains validity only for a carefully selected subset of measurement samples. This selection is achieved by a combination of three actions.

1. First, the FEM identifies moments during the discharge process when the current values are low. This condition minimizes voltage drops caused by the internal impedance, ensuring a more accurate representation of the open-circuit voltage. This requirement is imposed by the threshold condition in (17):

$$|I_t| \leq I_{thresh}, \tag{17}$$

where I_t is the current measurement and I_{thresh} is the current low pass threshold.

2. In addition, since the battery's internal impedance might present transients, a second condition must be considered in series to the first one. The FEM checks if the low current measurements are consistent for a given amount of time to minimize the effect of voltage transients due to impedance. The expression in Equation (18) represents the logical operator that indicates that a voltage measurement is close to its corresponding V_{oc} and can be formulated as:

$$T_{consistent} \geq T_{thresh}, \tag{18}$$

where $T_{consistent}$ is a counter that indicates how much time with low current has passed, and T_{thresh} represents the second control parameters that regulate how permissive the selector is. The combined condition can be formulated as:

$$(|I_t| \leq I_{thresh}) \wedge (T_{thresh} \leq T_{consistent}) \implies V_t \sim v_{oc}(SoC_t), \tag{19}$$

By applying this logical condition to the voltage and current signals, the FEM generates a set of observations $\mathcal{D}$.

3. Despite the low stable currents, they might still alter the approximation of V_{oc} when multiplied by the internal resistance. For this reason, an internal impedance compensation module is added to the flow as explained in Section 3.2.2. The application of this additional module adjusts the observation $\mathcal{D}$ to $\mathcal{D}_{comp}$.

In addition, the FEM logs the first qualifying voltage measurement that serves as a reference point for the battery capacity estimation process, called the reference voltage. The reference voltage is the initial V_o approximation that meets both conditions. V_o is then converted to an initial State of Charge SoC_0 using the inverse function of Equation (20). This transformation is possible as the function $v_{oc}(\cdot)$, presented in Equation (2), is bijective. Although the inverse function of $v_{oc}(\cdot)$ exists, it does not have an analytical form, necessitating an approximation. The proposed approximation, which maps v_{oc} measurements to their equivalent SoC, is given by the following:

$$SoC(V_{oc}) = \frac{x_2}{1 + e^{-x_1(V_{oc} - x_0)}} + (a \cdot V_{oc} + b)^3 + c, \quad 0 \leq SoC(V_{oc}) \leq 1, \tag{20}$$

where x_0, x_1, x_2, a, b, and c are adjustable parameters that must be tuned to minimize the error between the actual inverse function.

Finally, the FEM also computes the extracted charge between voltage detections. These values are computed by integrating the current over time. This process can be expressed as:

$$Q_{\text{ext}}(k) = \int_0^k I_t \, dt \approx \sum_{i=0}^{k} I(i) \cdot \Delta t_i \, , \tag{21}$$

where I_t corresponds to the current signal and Δt_i the sampling period between measurements.

When the FEM is applied to the monitored battery current and voltage signals, the algorithm generates a feature set formalized as:

$$\mathcal{D} = \{\mathcal{D}^{(i)} = (SoC_0^{(i)}, \Delta Q_{ext}^{(i)}, V_{oc}^{(i)}) | i \in \{1, 2, 3, \ldots, N\}\} \tag{22}$$

where $SoC_0^{(i)}$ corresponds to the initial SoC associated with the i-th selected measurement (computed with Equation (20)), $\Delta Q_{ext}^{(i)}$ is the extracted charge between the initial and the i-th selected observation, and $V_{oc}^{(i)}$ is the i-th selected voltage measurement.

As previously stated, despite the selection of a low stable current, the impact of the voltage drop due to the internal resistance could still be relevant. It is hence important to further compensate the observation $\mathcal{D}$ with the Resistance Effect Compensation Module.

3.2.2. Resistance Effect Compensation

In this work, the battery ohmic losses are modeled through a internal resistance R_{int}, which plays the role of the Equivalent Series Resistance (ESR) in the Thévenin model. This resistance can be characterized based on the data provided in [58], it is possible to fit a third degree polynomial to describe internal resistance as a function of SoC, where the polynomial coefficients are $P_0 = 0.1317$, $P_1 = -0.05083$, $P_2 = -0.2579$, and $P_3 = 0.3084$. The relationship can be expressed by:

$$R_{\text{int}}(\text{SoC}; I = 5A) = P_0 + P_1 \cdot \text{SOC} + P_2 \cdot \text{SOC}^2 + P_3 \cdot \text{SOC}^3 \tag{23}$$

This polynomial function represents a case in which the battery operates at a constant discharge current. However, actual operational conditions might present varied discharge current profiles. Therefore, we propose adjusting the polynomial using historical operational voltage and current data. The adjustment of Equation (23) involves an affine transformation, as illustrated in Equation (24), where θ represents adjustable parameters.

$$R_{\text{int}}(\text{SOC}, I)_\theta = (P_0 + P_1 \cdot \text{SOC} + P_2 \cdot \text{SOC}^2 + P_3 \cdot \text{SOC}^3) \cdot \theta_1 + \theta_2 \tag{24}$$

Then, Equation (25) describes how to refine the detected V_{oc} measurements by considering the ohmic effects of impedance at low currents. Note that $R_{int_k}(SoC, I_k)_\theta$ in Equation (25) corresponds to the proposed parameterization of the internal resistance.

$$v_{oc}(SoC_k) = V_k + I_k \cdot R_{int_k}(SoC_k, I_k)_\theta, \tag{25}$$

This compensation scheme is applied to the voltage detections of the feature set presented in Equation (22), to diminish the effect of internal resistance further. The application of this procedure generates the compensated feature dataset given by:

$$\mathcal{D}_{comp} = \left\{\mathcal{D}^{(i)} = \left(SoC_0^{(i)}, \Delta Q_{\text{ext}}^{(i)}, V_{oc}^{(i)} + I^{(i)} \cdot R_{int_k}(SoC(V_{oc}^{(i)}), I^{(i)})_\theta\right) \mid i \in \{1, 2, \ldots, N\}\right\} \tag{26}$$

$$= \left\{\mathcal{D}^{(i)} = \left(SoC_0^{(i)}, \Delta Q_{\text{ext}}^{(i)}, \hat{V}_{oc}^{(i)}\right) \mid i \in \{1, 2, \ldots, N\}\right\}. \tag{27}$$

where $I^{(i)}$ represents the current associated with the i-th detection, and $\hat{V}_{oc}^{(i)}$ represents the i-th compensated voltage detection of the compensated feature dataset $\mathcal{D}_{comp}$. After compensating for resistive effects, no further processing is necessary.

It is noteworthy to mention that the proposed polynomial model for the internal impedance corresponds to a simplified empirical approximation. Indeed, internal impedance depends on several factors, including temperature, hysteresis, and manufacturing defects. Since these factors are not being taken as inputs to the model, errors in the characterization of internal impedance are expected.

However, these errors do not compromise the validity of the proposed methodology. First, the polynomial model is calibrated for the specific cell type under study using historical data from similar cells. Second, as shown in Equation (25), errors in the term $R_{int_k}(SoC_k, I_k)_\theta$ are multiplied by I_k, whose magnitude is inherently small during the operating conditions considered by this methodology. Consequently, any approximation error in the internal resistance remains bounded. Finally, residual errors introduced by this simplification are accounted for in the observation equation of the upstream particle filter (PF) as observation noise. Therefore, the influence of inaccuracies in $R_{int_k}(SoC_k, I_k)_\theta$ on the overall performance of the proposed methodology can be considered negligible.

3.2.3. Maximum Likelihood Estimation Module

Once the features are extracted, the battery capacity parameter can be estimated by maximum likelihood. To construct the likelihood function of the battery capacity, the following assumptions are considered:

- During the feature extraction, capacity degradation is neglected.
- The extracted features are conditionally identically independent and identically distributed.
- The extracted open voltage features are affected by Gaussian additive noise with some variance σ_v^2.

In particular, the latter assumption is grounded in established practices. Modeling voltage-related features as the true cell voltage corrupted by additive, zero-mean Gaussian noise is in line with established practices in Li-ion battery state estimation, where terminal-voltage measurements acquired through the sensing chain are typically affected by small, approximately white noise perturbations. Under this setting, the Gaussian model provides a convenient and well-posed likelihood for machine learning or Bayesian estimators as proposed by [36,59–61]. In applications where the sensor noise is manifestly non-Gaussian, the same estimation framework can be reformulated with a more general noise model, at the price of increased complexity. With the compensated set of extracted features presented in Equation (27) and a previously adjusted battery model, the V_{oc} can be approximated for every observation, following Equation (28). Note that this voltage prediction depends on the battery capacity C. This allows the creation of an error metric assumed to be normally distributed. Equation (29) shows the final error metric associated with the collected feature vector i.

$$\hat{V}_{oc}^{(i)}(C, SoC_0^{(i)}, \Delta Q_{\text{ext}}^{(i)}) = v_{oc}\left(SoC_0^{(i)} - \frac{\Delta Q_{\text{ext}}^{(i)}}{C}\right) \tag{28}$$

$$\begin{aligned}\epsilon(C, \mathcal{D}^{(i)}) &= \hat{V}_{oc}^{(i)} - v_{oc}^{(i)}(C, V_0^{(i)}) \\ &= \hat{V}_{oc}^{(i)} - v_{oc}\left(SoC_0^{(i)} - \frac{\Delta Q_{\text{ext}}^{(i)}}{C}\right)\end{aligned} \tag{29}$$

When combining the above-mentioned assumptions, the likelihood function of a set of observations follows:

$$\mathcal{L}(C; \mathcal{D}) = \prod_{i=1}^{N} \mathcal{L}(C; \mathcal{D}^{(i)}) \tag{30}$$

$$= \prod_{i=1}^{N} \frac{1}{\sigma_v \sqrt{2\pi}} e^{-\frac{1}{2}(\frac{\epsilon(C,\mathcal{D}^{(i)})}{\sigma_v})^2} \tag{31}$$

$$\propto e^{\sum_{i=1}^{N} -\frac{1}{2}(\frac{\epsilon(C,\mathcal{D}^{(i)})}{\sigma_v})^2} \tag{32}$$

Since we are interested in the most probable estimation of the battery capacity, the estimated value $\hat{C}$ is the one that maximizes the constructed likelihood. To ease the computations, we use the log-likelihood instead of the likelihood function. Additionally, the multiplicative factors can be ignored during the maximization process. The optimization problem is described by Equation (36).

$$\hat{C} = \arg\max_{C} log(\mathcal{L}(C; \mathcal{D})) \tag{33}$$

$$= \arg\max_{C} log(e^{\sum_{i=1}^{N} -\frac{1}{2}(\frac{\epsilon(Q,\mathcal{D}^{(i)})}{\sigma_v})^2}) \tag{34}$$

$$= \arg\min_{C} \sum_{i=1}^{N} [\epsilon(C, \mathcal{D}^{(i)})]^2 \tag{35}$$

$$= \arg\min_{C} \sum_{i=1}^{N} [\hat{V}_{oc}^{(i)} - v_{oc}(SoC_0^{(i)} - \frac{\Delta Q_{ext}^{(i)}}{C})]^2 \tag{36}$$

The MLE employs a gradient descent algorithm to generate capacity estimates from operational data. When the optimization problem is solved, the MLE Module outputs an estimation of the battery capacity. Although this estimator can be used as a virtual capacity sensor, the accuracy of the estimates is based solidly on the collected data, which means that short discharges might produce biased estimates. In addition, noisy sensors may produce unreliable data points and result in outliers. The MLE-PF framework merges the MLE with a PF loop to overcome this issue. As a result, the MLE acts as a virtual sensor of battery capacity, whose outputs are integrated as observations by the PF in a higher-level loop. These MLE observations are used to update the particle weights in order to compute the posterior PDF. In this way, inaccurate or incomplete estimates can be corrected using prior information from the degradation model. In particular, the PF algorithm uses a Gaussian likelihood function, which gives an estimate of the battery capacity in a Bayesian fashion. The modular nature of the framework allows for the integration of any degradation model within the Particle Filter. In this work, we employ the SoC estimator proposed in [37], along with the degradation model formulated in [40]. Note that once a posterior distribution of battery capacity is generated, its expected value is used to update the SoC estimator, which uses the battery capacity as a normalizing constant to calculate the SoC. This feedback loop allows for a simultaneous assessment of health and charge, allowing complete battery characterization during operation.

PF and estimator parameters should be adjusted using an initial set of training cycles. The procedure presented in [37] allows the parameter adjustment of the battery model using the operational data of the EV. The variance of the uncertainty of the model process can be inferred using a small batch of degradation tests. However, other outer feedback correction loops can be used to dynamically adjust this parameter during operation [62]. Finally, the likelihood variance depends only on the voltage sensor noise variance; hence, it can be consulted with the vehicle manufacturer or inferred from operational data. In the case study, this calibration step is performed only on cell W9, which plays the role of a training cell. The resulting inverse-OCV/ESR parameters and PF noise variances are then kept fixed when evaluating cells W4, W8, and W10, thus avoiding any leakage of test information into the tuning stage.

The proposed method uses a PF algorithm since the sequential estimation problem is characterized by non-linear degradation dynamics. Although both methods are computationally expensive, this characteristic is not limiting for an on-board application, since capacity estimates and PF updates are triggered once an equivalent discharge cycle is completed. This dynamics implies that the execution time of the algorithm is much shorter than the battery degradation rates.

4. Case Study

Dataset Description

To validate the MLE-PF framework proposed in this work, we used the publicly available lithium-ion battery aging dataset provided by the Stanford University and released by the MIT–Toyota Research Center [34]. This dataset documents the degradation of ten commercial INR21700-M50T lithium-ion cells (NMC–graphite/silicon chemistry, nominal capacity of 4850 mAh) cycled over a 23-month period under realistic use scenarios. It encompasses multiple experimental variables and reproduces realistic EV profiles at a controlled temperature of 25 °C, with power demands scaled to the cell's operational range.

Each cycle consists of full charge–discharge sequence between 100% and 20% of the SoC. Charging is performed through a two-phase Constant-Current Constant-Voltage (CCCV) protocol using C-rates between $C/4$ and 3C for different cells until 100% of the SoC is reached and the charging current falls below 50 mA. This is followed by a constant-current discharge at $C/4$ down to approximately 80% SoC and subsequently by the realistic conduction profile Urban Dynamometer Driving Schedule (UDDS) discharge to reach 20% SoC.

Diagnostic testing is performed periodically, including capacity testing using Hybrid Pulse Power Characterization (HPPC) and EIS. By the end of the campaign, cells reached between 80% and 91% of their initial capacity. Detailed information on the voltage, current, charge, and discharge capacities, time, and step-index measurements for each cell cycle are recorded. The following variables have been added or modified by the authors in a pre-processing phase:

- SoC: calculated for each experiment by integrating the current over time. This value is normalized with the SoC variation limits (20% and 100%).
- SoH: It is important to note that this dataset does not directly provide the SoH values; instead, SoH has been derived from capacity measurements. Additionally, the current

capacity data is not provided for each cycle but is available only at specific assessment intervals. To establish a continuous SoH trend with values at each cycle, linear interpolation has been employed to estimate capacity (and consequently SoH) between the reported data points. This interpolated data will serve as the reference for the cell SoH. This interpolated data will serve as the reference for the cell SoH. However, in Section 5 the MLE–PF estimates are also compared against the non-interpolated ground-truth capacity values at diagnostic cycles, and these errors are reported as the primary performance metrics.

Among the ten cells, four have been selected for this study due to their higher degree of degradation, allowing a thorough evaluation of the performance of the proposed model. As a result, the pre-processing step was applied to cells W4, W8, W9, and W10. Experimental data from cell W9 were used exclusively to tune the parameters and hyperparameters of the proposed methodology; consequently, cell W9 is not used to validate it. In contrast, the remaining three cells have been thoroughly analyzed, and the results are provided in Section 5.

The specific charging conditions of each cell, ambient temperature, and diagnostics performed, including the number of charge-discharge cycles each cell had completed at the time of each diagnostic assessment, are detailed in Table 1. It is crucial to note that the frequency of diagnostic tests that evaluate the SoH of cells is not uniform; that is, they are not performed after a constant predetermined number of cycles. In some cases, these tests are performed every 25 cycles; in others, they may be conducted every 48 cycles. As a result, the initial capacity value is defined as the one reported in the first cycle immediately after a diagnostic test, and the final value corresponds to the capacity measured in the last cycle before the next diagnostic test. The capacity test is carried out by first bringing the cell to 100% SOC and then discharging it with a current equal to one-twentieth of its nominal capacity. Using such a very low current keeps over-potential losses practically zero, so the voltage profile follows the open-circuit curve, and the charge removed can be taken as an accurate measure of the cell's actual capacity.

Table 1. Operating conditions of the studied cells. For each diagnostic assessment (D1–D9), the table reports how many charge-discharge cycles each cell had completed at the time of the corresponding diagnostic test. Adapted from [63].

Cell	T [°C]	Charge C-Rate	D1	D2	D3	D4	D5	D6	D7	D8	D9
W4	23	C/4	0	25	75	123	132	159	176	179	N/A
W8	23	C/2	0	25	75	125	148	150	151	157	185
W9	23	1C	0	25	75	122	144	145	146	150	179
W10	23	3C	0	25	75	122	146	148	151	159	188

5. Results

In this section, the results regarding cells W4, W8, and W10, selected due to their higher degree of degradation, are reported. In particular, we present the MLE-PF SoH estimates along with the ground truth values reported in the Stanford dataset. To quantify how well the model follows the actual degradation dynamics, we computed the Mean Absolute Error (MAE) for each cell in a box plot visualization. In addition, we performed a sensitivity analysis of the FEM with respect to the low-current threshold I_{th} and the minimum low-current duration T_{th}, by evaluating how the number of detected pseudo-open-circuit voltage points N_{OCV} varies across different parameter combinations. Furthermore, we analyze how the SoH estimation error (in terms of MAE) changes as a function of different (I_{th}, T_{th}) combinations to assess the robustness of the proposed MLE-PF framework.

5.1. Sensitivity Analysis of Low-Current Detection Parameters

The FEM relies on two hyperparameters that determine when a voltage sample can be treated as a pseudo-open-circuit voltage V_{ocp}: the low-current threshold I_{th} and the minimum low-current duration T_{th}, which correspond to the conditions in (19). To assess the robustness of the proposed methodology with respect to these parameters, we performed a sensitivity study on cell W4 of the Stanford dataset. For a grid of $(I_{\text{th}}, T_{\text{th}})$ combinations, we counted the total number of pseudo-OCV points, denoted by N_{OCV}, detected across all available work cycles using the same driving profile and data preprocessing described in Section 4.

Figure 4 indicates that different $(I_{\text{th}}, T_{\text{th}})$ combinations yield a comparable number of detected pseudo-OCV points N_{OCV}. For this reason, we restricted the detailed error analysis to the representative settings reported in Table 2, instead of exhaustively evaluating every pair in the grid.

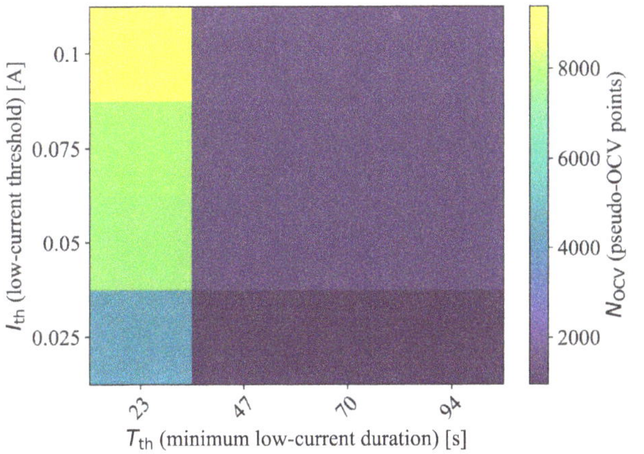

Figure 4. Sensitivity of the detected pseudo-OCV points to the low-current detection parameters. The heat map shows the total number of pseudo-OCV points N_{OCV} as a function of the current threshold I_{th} and the minimum low-current duration T_{th} for cell W4.

Table 2. Summary of sensitivity analysis results for cell W4 under different combinations of current and time thresholds.

I_{th} [A]	T_{th} [s]	Interpolated MAE [%]	MAE at Diagnostic Tests [%]	N_{OCV}
0.025	***47***	***0.235***	***0.349***	***1026***
0.050	47	0.480	0.420	1512
0.025	23	0.573	0.487	4270
0.050	23	0.507	0.503	7835
0.075	23	0.937	1.040	7927
0.100	23	0.850	1.020	9387

Bold italics indicate the best-performing parameter combination in the sensitivity analysis.

Table 2 reports, for each $(I_{\text{th}}, T_{\text{th}})$ combination, three metrics: the Interpolated MAE, computed over all work cycles; the MAE at diagnostic tests, evaluated only at the diagnostic cycles where reference capacity measurements are available; and the total number of detected pseudo-OCV points N_{OCV}. In both cases, the SoH estimation error is obtained by comparing the SoH estimated by the proposed MLE–PF framework against the reference SoH derived from the ground-truth capacity measurements in the Stanford dataset. From Table 2, the lowest MAE is obtained for $I_{\text{th}} = 0.025$ A and $T_{\text{th}} = 47$ s, both for the interpolated and non-interpolated errors. This combination corresponds to the smallest current threshold and the longest low-current duration among the tested cases, which is consistent with the physical intuition behind the Feature Extraction Module: as the current decreases and the low-current window becomes longer, the measured terminal voltage has more time to relax and becomes closer to the true OCV curve.

It is also worth noting that this best-performing configuration is associated with the smallest number of detected pseudo-OCV points within the tested set. This suggests that, at least for cell W4, the quality of the selected points—i.e., how close they are to the underlying OCV response—is more important for the MLE performance than simply maximizing N_{OCV}. Based on this trade-off, the pair $I_{\mathrm{th}} = 0.025$ A and $T_{\mathrm{th}} = 47$ s was adopted in the following experiments.

5.2. SoH Estimation Results

Figures 5–8 show the SoH estimation results obtained with the proposed MLE–PF algorithm for cells W4, W8, and W10. Each figure includes the MLE–PF output, the ground-truth reference, and the 5–95 % confidence interval estimated from the particle distribution. Orange markers indicate the estimated SoH values at diagnostic cycles, while the lower panel in each figure presents the corresponding absolute error evolution over all cycles.

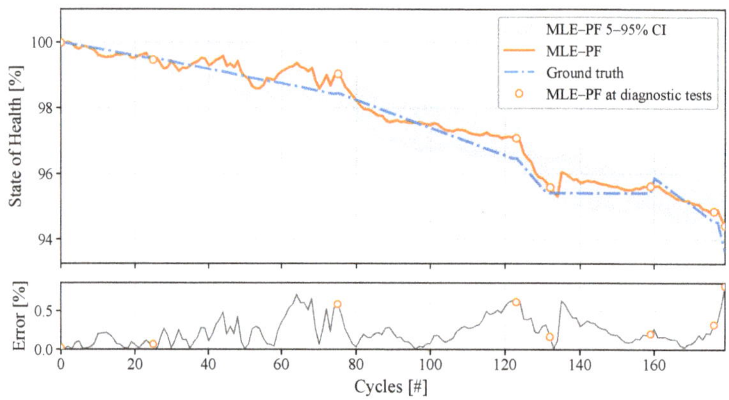

Figure 5. SoH estimation results for cell W4 showing the MLE–PF output, the ground-truth reference, and the posterior 5–95% credible interval (nominal 90% level), together with the error evolution across cycles (# denotes the cycle index).

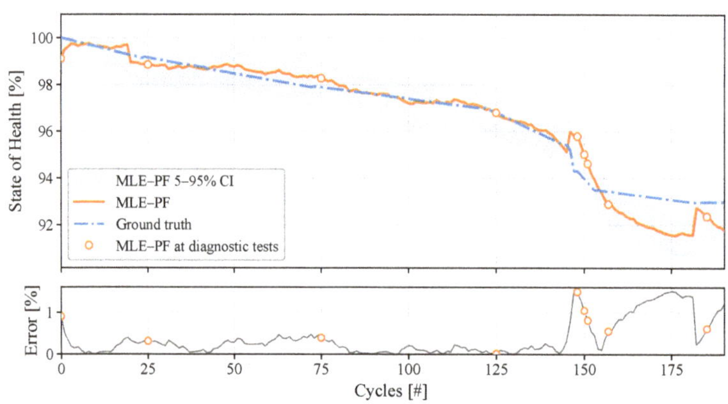

Figure 6. SoH estimation results for cell W8 showing the MLE–PF output, the ground-truth reference, and the posterior 5–95% credible interval (nominal 90% level), along with the error evolution across cycles. (# denotes the cycle index).

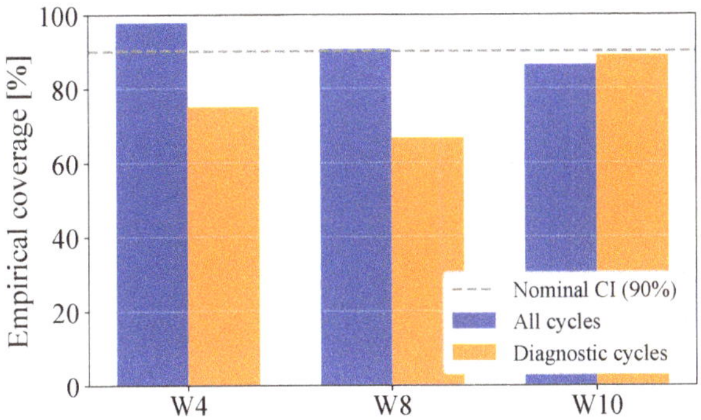

Figure 7. Empirical coverage of the 5–95% credibility interval for cells W4, W8, and W10. The dashed line indicates the nominal 90% level.

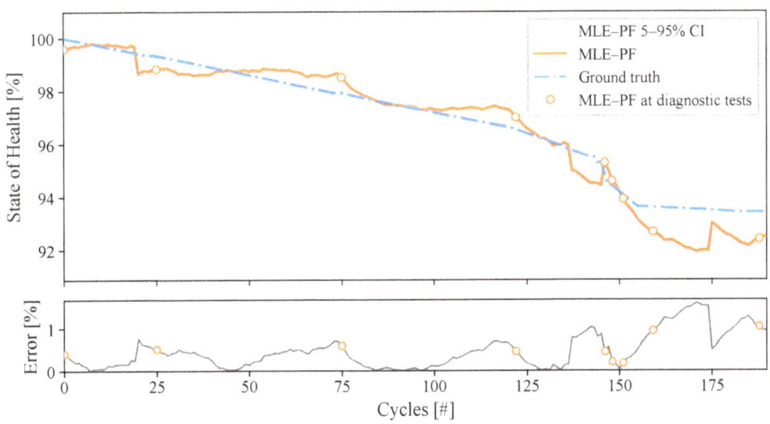

Figure 8. SoH estimation results for cell W10, showing the MLE–PF output, the ground-truth reference, and the posterior 5–95% credible interval (nominal 90% level), along with the error evolution across cycles (# denotes the cycle index).

Figure 5 shows the results obtained for cell W4, where the Stanford dataset indicates an overall approximately 6% degradation in SoH across the full cycling range. The plot demonstrates that the proposed MLE-PF model effectively tracks the cell's degradation throughout its operational life. Notably, the point-wise error remains comfortably below 0.5%, and the overall MAE is 0.235% when computed over all work cycles and 0.349% at diagnostic test points (see Table 3), indicating consistent accuracy and stable convergence of the estimator. The 5–95 % confidence interval remains narrow over most of the test, confirming that the particle distribution captures the expected variability while preserving a low estimation uncertainty.

Table 3. SoH estimation MAE. The interpolated MAE is computed over all work cycles using the linearly interpolated ground-truth values, while the diagnostic MAE is calculated only at diagnostic test cycles.

Cell	Interpolated MAE [%]	MAE at Diagnostic Tests [%]	Diagnostic Points Used
W4	0.235	0.349	8
W8	0.369	0.686	9
W10	0.465	0.517	9

For cell W8, the results shown in Figure 6 depict a similar scenario, with an interpolated MAE of 0.369% and a MAE at diagnostic tests of 0.686% (Table 3), closely following the actual degradation trend, although the error increases near the final cycles after the drop around 145–155.

Finally, the results for cell W10, presented in Figure 8, exhibit comparable performance in capturing degradation dynamics, with a resulting interpolated MAE of 0.465% and a MAE at diagnostic tests of 0.517% (Table 3); the late-cycle deviations remain bounded by the credibility band.

Overall, the results indicate stable precision and support the effectiveness of the MLE–PF in predicting the SoH trend across cells. As summarized in Table 3, with W4 achieving the lowest overall error, W8 showing a slight increase during the final degradation stage, and W10 maintaining a consistent tracking of the reference trend. Figure 9 illustrates these results by representing the estimation error distributions of cells W4, W8, and W10 with box plots, reflecting the same relative behavior reported in Table 3. Figure 7 shows that the empirical coverage of the 5–95% credibility band is close to the nominal 90% level when all cycles are considered. At diagnostic cycles the coverage is lower for W4 and W8, while W10 is closer to the nominal value. Since all three cells have only a small number of diagnostic measurements, these percentages are sensitive to a few mismatches; in a realistic deployment with more diagnostic data available, the empirical coverage at those points would be expected to fluctuate less and to concentrate around the nominal level.

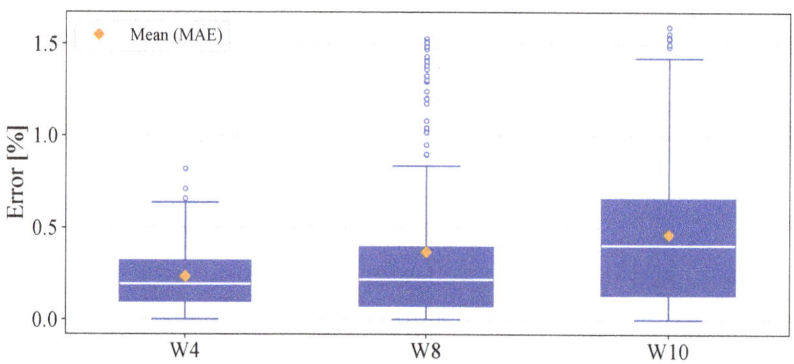

Figure 9. Distribution of SoH estimation errors for cells W4, W8, and W10.

6. Discussion

The results obtained from the performed experiments confirm the accuracy of the proposed methodology for online lithium-ion cell SoH estimation. The methodology considers a feature selection module that saves the most significant measurements during operation to create a feature set that is later employed to generate MLE capacity estimates. These estimates are later filtered to suppress possible artifacts when the data presents anomalies. In a practical implementation, a device should capture current and voltage measurements from the CAN bus and then process them to generate capacity estimates and filter updates once equivalent cycles are completed. Note that this latter processing can be performed with cloud computing if the computational capacity of the onboard device is limited. These findings are consistent across all cells and align with the error summary presented in Table 3.

Beyond these baseline results, the sensitivity analysis reported in Section 5.1 (Figure 4 and Table 2) indicates that the MLE-PF framework is robust to different combinations of the low-current threshold I_{th} and the minimum low-current duration T_{th}, with similar MAE values for most of the tested settings. At the same time, the smallest errors are obtained when pseudo-OCV points are extracted under the most restrictive conditions, that is, using the lowest current threshold and the longest dwell time, which is consistent with the idea that lower currents and longer relaxation intervals produce voltage samples closer to the true OCV. These two hyperparameters were selected for the sensitivity study because they can be directly tuned in a practical implementation, whereas the uncertainties associated with the ESR compensation are absorbed into the process and measurement noise of the MLE-PF framework and are identified offline under the nearly constant-temperature conditions of the Stanford dataset. A systematic assessment of the sensitivity of ESR model parameters to temperature variations remains an open topic for future investigation.

From an application point of view, the outcome of this online SoH estimation procedure can serve many purposes during the lifetime of the vehicle. For instance, since autonomy directly depends on the battery capacity, having SoH estimates can drastically enhance precision in autonomy prognosis algorithms, especially at the end of the battery life cycle.

The high versatility and scalability of the framework lie in its modularity, which enables the application of the framework on different types of batteries with different chemistries, after an offline parameter tuning phase. In particular, in this case, the implementation of Thévenin-equivalent battery models can reproduce the electrical behavior of batteries, packs, and cells from different chemistries and configurations.

In addition to these benefits, our approach differs from existing methods in at least two other significant aspects. First, it adapts to the usage of a real electric vehicle without a specific discharge profile. Several model-based methods handle degradation under different C-rates but still assume stationary or fixed operating profiles throughout each discharge cycle [64]. These assumptions are impractical in an EV, where driving patterns vary daily. In contrast, we only require samples in the operational data where the current is sufficiently low to treat the measured voltage as an approximation of the open-circuit voltage. This requirement is easy to meet during regular driving, as vehicles often remain near zero current while coasting or at low power demand. Thus, we estimate capacity by collecting these naturally occurring measurements over one or more trips, which makes the method suitable for everyday EV operation.

Second, the proposed algorithm needs a small set of parameters that can be extracted from a few conventional tests or short operational campaigns. The Thévenin-equivalent model [65] demands limited information, unlike machine learning models based on large labeled data sets and long training times [17].

In our method, parameter tuning involves recording voltage and current data from ordinary vehicle usage, fitting the open-circuit voltage curve with a few parameters, and characterizing internal resistance with a straightforward polynomial. This procedure alleviates the computational demands associated with training complex neural networks or gathering extensive aging data, while also improving explainability, an pivotal aspect in safety-critical scenarios. Because the model structure is simple and the number of parameters is small, the estimator can be executed with limited processing capability and does not require laboratory-grade calibration. This simplicity is particularly advantageous in situations where the initial state of the battery is uncertain or only roughly defined, which is unfortunately often encountered in practical EV applications. Under those conditions, highly complex models tend to be sensitive to initialization or to gaps in the training data, whereas a low-dimensional parametric description can still produce reasonable intermediate estimates as soon as real measurements become available. Here the PF plays a key role: by propagating a set of hypotheses and weighting them with actual observations, the PF progressively corrects the initial uncertainty and steers the estimate toward the behavior observed in the vehicle. This is particularly relevant in EV operation, where charge and discharge cycles are rarely complete and the information content of the data is lower than in laboratory cycling: even with these fragmented, opportunistic profiles, the PF can converge towards the true SoH trajectory after a few updates. Consequently, the proposed estimator remains easy to implement on low-cost hardware, adapts quickly to practical EV usage conditions, and converges robustly to accurate SoH estimates in real deployments.

7. Conclusions

This work presents a practical and scalable approach for real-time SoH estimation of lithium-ion batteries in electric vehicles. It operates using standard operational data, eliminating the need for predefined discharge profiles or controlled degradation tests. This makes it well-suited for seamless integration into routine EV operations. Moreover, the approach is inherently scalable across different lithium-ion cells, and even for modules and battery configurations, as it does not depend on degradation datasets for parameter tuning. By leveraging a Thévenin-equivalent circuit model, the parameter estimation process remains computationally efficient and adaptable, avoiding the complexity typically associated with machine learning-based methods. Across the case study, the proposed method tracks SoH with mean absolute errors below 1%, using only on-board voltage and current signals.

A critical aspect of this method is its reliance on the OCV model, which provides a detailed understanding of battery behavior under various operational conditions. Coupling this OCV model with the MLE framework enables accurate capacity estimation, even when starting from an arbitrary SoC. This capability is crucial in real-time applications, where battery conditions often vary significantly and unpredictably. In addition, the inclusion of PF adds a layer of robustness by accounting for uncertainties in the measurement process and the inherent variability of battery performance. The synergy between MLE's statistical basis and PF's adaptive framework corrects deviations in SoH estimates while ensuring that the predicted capacity remains closely aligned with the degradation dynamics observed in the dataset.

The sensitivity analysis of the low-current detection parameters further indicates that accurate SoH tracking can be achieved with a relatively small number of high-quality pseudo-OCV points, provided that the current and time thresholds are chosen to favor measurements close to true open-circuit conditions.

Taken together with the experimental results on cells W4, W8, and W10, these findings confirm the efficacy of the proposed method for estimating the SoH of lithium-ion batteries with standard operational profiles. By leveraging operational data, this method provides a noninvasive, efficient, and accurate tool to monitor battery degradation in real-time.

Overall, the method developed in this work offers a practical and scalable solution for real-time SoH monitoring of lithium-ion batteries in electric vehicles. Its application can significantly enhance the efficiency of battery management systems, contributing to a broader adoption of EVs by ensuring reliability and safety. Future work will focus on further refining the model parameters and exploring the integration of additional data sources, such as temperature and external load variations, to improve the accuracy and applicability of the SoH estimation process.

Author Contributions: Conceptualization, J.E.G.B., B.B.S., F.J.-M., D.T.-K., M.E.O., A.P. and J.F.S.; Methodology, J.E.G.B., B.B.S. and F.J.-M.; Software, J.E.G.B., B.B.S. and F.J.-M.; Validation, J.E.G.B., B.B.S. and D.T.-K.; Formal analysis, J.E.G.B., B.B.S. and L.B.; Investigation, J.E.G.B., B.B.S., L.B. and B.M.; Resources, J.E.G.B., L.B., D.T.-K. and A.P.; Data curation, J.E.G.B., B.B.S. and B.M.; Writing—original draft, J.E.G.B., B.B.S., L.B., B.M., F.J.-M., M.E.O. and A.P.; Writing—review & editing, J.E.G.B., L.B., B.M., F.J.-M., D.T.-K., M.E.O., A.P. and J.F.S.; Visualization, J.E.G.B., B.B.S. and B.M.; Supervision,

F.J.-M., M.E.O., A.P. and J.F.S.; Project administration, L.B., M.E.O., A.P. and J.F.S.; Funding acquisition, M.E.O., A.P. and J.F.S. All authors have read and agreed to the published version of the manuscript.

Funding: This work was supported in part by ANID FONDECYT 1250036, ANID FONDECYT 11231148, Advanced Center for Electrical and Electronic Engineering, ANID Basal Project CIA250006. The work of Jorge E. García Bustos has been supported by ANID-PFCHA/Doctorado Nacional/2022-21221213. Leonardo Baldo: This publication is part of the project PNRR-NGEU, which has received funding from the MUR–DM 352/2022. The work of Aramis Perez was supported by the University of Costa Rica under research projects 322-C1-467.

Data Availability Statement: The data used in this study were derived from a publicly available dataset [https://doi.org/10.1016/j.dib.2022.107995], but the processed results generated during the analysis are not publicly available. These derived data can be provided by the authors upon reasonable request.

Conflicts of Interest: The authors declare no conflicts of interest.

Nomenclature

The following abbreviations are used in this manuscript:

Abbreviation	Definition
EV	Electric Vehicle
LIB	Lithium-Ion Batteries
DoD	Depth of Discharge
SoH	State of Health
EIS	Electrochemical Impedance Spectroscopy
BMS	Battery Management System
PF	Particle Filter
SoC	State of Charge
RUL	Remaining Useful Life
SoH	State of Health
PHM	Prognostic and Health Management
CAN	Controller Area Network
MLE	Maximum Likelihood Estimation
MLE-PF	Maximum Likelihood Estimation-Particle Filter
BF	Bayesian Filter
OCV	Open-Circuit Voltage
SIS	Sequential Importance Sampling
PDF	Probability Density Function
FEM	Feature Extraction Module
CCCV	Constant-Current Constant-Voltage
UDDS	Urban Dynamometer Driving Schedule
HPPC	Hybrid Pulse Power Characterization
t	time step index
k	Work cycle index
I	Current
Δt	Sampling time
w	SoC process noise
V_{oc}	Open-circuit voltage
V_{oc_p}	Pseudo-open-circuit voltage
C	Battery capacity parameter (used in the MLE module)
C_k	Battery capacity at work cycle k
C_0	Initial (nominal) battery capacity
η	Efficiency factor
v	Capacity process noise

L	Likelihood function
θ	MLE parameters
$f(x_i;\theta)$	Probability density function of the data x_i given the parameters θ
T	Low current counter
$\mathcal{D}$	Set of observations
R_{int}	Battery internal resistance
Q_{ext}	Extracted charge
I_{th}	Low-current threshold used in the FEM
T_{th}	Minimum low-current duration used in the FEM
N_{OCV}	Total number of detected pseudo-OCV points
ESR	Equivalent Series Resistance (internal resistance of the cell)

References

1. Velev, B.; Djudzhev, B.; Dimitrov, V.; Hinov, N. Comparative Analysis of Lithium-Ion Batteries for Urban Electric/Hybrid Electric Vehicles. *Batteries* **2024**, *10*, 186. [CrossRef]
2. Chebiam, R.; Kannan, A.; Prado, F.; Manthiram, A. Comparison of the chemical stability of the high energy density cathodes of lithium-ion batteries. *Electrochem. Commun.* **2001**, *3*, 624–627. [CrossRef]
3. Edge, J.S.; O'Kane, S.; Prosser, R.; Kirkaldy, N.D.; Patel, A.N.; Hales, A.; Ghosh, A.; Ai, W.; Chen, J.; Yang, J.; et al. Lithium ion battery degradation: What you need to know. *Phys. Chem. Chem. Phys.* **2021**, *23*, 8200–8221. [CrossRef]
4. Wang, Q.; Ping, P.; Zhao, X.; Chu, G.; Sun, J.; Chen, C. Thermal runaway caused fire and explosion of lithium ion battery. *J. Power Sources* **2012**, *208*, 210–224. [CrossRef]
5. Saw, L.H.; Ye, Y.; Tay, A.A. Integration issues of lithium-ion battery into electric vehicles battery pack. *J. Clean. Prod.* **2016**, *113*, 1032–1045. [CrossRef]
6. Soyoye, B.D.; Bhattacharya, I.; Anthony Dhason, M.V.; Banik, T. State of Charge and State of Health Estimation in Electric Vehicles: Challenges, Approaches and Future Directions. *Batteries* **2025**, *11*, 32. [CrossRef]
7. Zhou, Y.; Huang, M.; Chen, Y.; Tao, Y. A novel health indicator for on-line lithium-ion batteries remaining useful life prediction. *J. Power Sources* **2016**, *321*, 1–10. [CrossRef]
8. Lu, J.; Xiong, R.; Tian, J.; Wang, C.; Sun, F. Deep learning to estimate lithium-ion battery state of health without additional degradation experiments. *Nat. Commun.* **2023**, *14*, 2760. [CrossRef]
9. Lipu, M.H.; Hannan, M.; Hussain, A.; Hoque, M.; Ker, P.J.; Saad, M.; Ayob, A. A review of state of health and remaining useful life estimation methods for lithium-ion battery in electric vehicles: Challenges and recommendations. *J. Clean. Prod.* **2018**, *205*, 115–133. [CrossRef]
10. Sohaib, M.; Akram, A.S.; Choi, W. Analysis of Aging and Degradation in Lithium Batteries Using Distribution of Relaxation Time. *Batteries* **2025**, *11*, 34. [CrossRef]
11. Yan, Z.; Wang, X.; Wei, X.; Dai, H.; Liu, L. State-of-Health Estimation of LiFePO4 Batteries via High-Frequency EIS and Feature-Optimized Random Forests. *Batteries* **2025**, *11*, 321. [CrossRef]
12. Jiang, B.; Zhu, J.; Wang, X.; Wei, X.; Shang, W.; Dai, H. A comparative study of different features extracted from electrochemical impedance spectroscopy in state of health estimation for lithium-ion batteries. *Appl. Energy* **2022**, *322*, 119502. [CrossRef]
13. Noura, N.; Boulon, L.; Jemeï, S. A review of battery state of health estimation methods: Hybrid electric vehicle challenges. *World Electr. Veh. J.* **2020**, *11*, 66. [CrossRef]
14. Nuroldayeva, G.; Serik, Y.; Adair, D.; Uzakbaiuly, B.; Bakenov, Z. State of Health Estimation Methods for Lithium-Ion Batteries. *Int. J. Energy Res.* **2023**, *2023*, 4297545. [CrossRef]
15. Hu, X.; Yuan, H.; Zou, C.; Li, Z.; Zhang, L. Co-Estimation of State of Charge and State of Health for Lithium-Ion Batteries Based on Fractional-Order Calculus. *IEEE Trans. Veh. Technol.* **2018**, *67*, 10319–10329. [CrossRef]
16. Shen, P.; Ouyang, M.; Lu, L.; Li, J.; Feng, X. The Co-estimation of State of Charge, State of Health, and State of Function for Lithium-Ion Batteries in Electric Vehicles. *IEEE Trans. Veh. Technol.* **2018**, *67*, 92–103. [CrossRef]
17. Li, K.; Wang, Y.; Chen, Z. A comparative study of battery state-of-health estimation based on empirical mode decomposition and neural network. *J. Energy Storage* **2022**, *54*, 105333. [CrossRef]
18. Ghosh, N.; Garg, A.; Panigrahi, B.; Kim, J. An Evolving Quantum Fuzzy Neural Network for online State-of-Health estimation of Li-ion cell. *Appl. Soft Comput.* **2023**, *143*, 110263. [CrossRef]
19. Feng, X.; Weng, C.; He, X.; Han, X.; Lu, L.; Ren, D.; Ouyang, M. Online State-of-Health Estimation for Li-Ion Battery Using Partial Charging Segment Based on Support Vector Machine. *IEEE Trans. Veh. Technol.* **2019**, *68*, 8583–8592. [CrossRef]
20. Etxandi-Santolaya, M.; Montes, T.; Casals, L.C.; Corchero, C.; Eichman, J. Data-Driven State of Health and Functionality Estimation for Electric Vehicle Batteries Based on Partial Charge Health Indicators. *IEEE Trans. Veh. Technol.* **2025**, *74*, 5321–5334. [CrossRef]

21. Zhang, Y.; Gu, P.; Duan, B.; Zhang, C. A hybrid data-driven method optimized by physical rules for online state collaborative estimation of lithium-ion batteries. *Energy* **2024**, *301*, 131710. [CrossRef]
22. Dong, H.; Mao, L.; Qu, K.; Zhao, J.; Li, F.; Jiang, L. State of Health Estimation and Remaining Useful Life Estimation for Li-ion Batteries Based on a Hybrid Kernel Function Relevance Vector Machine. *Int. J. Electrochem. Sci.* **2022**, *17*, 221135. [CrossRef]
23. Xiang, Y.; Fan, W.; Zhu, J.; Wei, X.; Dai, H. Semi-supervised deep learning for lithium-ion battery state-of-health estimation using dynamic discharge profiles. *Cell Rep. Phys. Sci.* **2024**, *5*, 101763. [CrossRef]
24. Wang, C.; Su, Y.; Ye, J.; Xu, P.; Xu, E.; Ouyang, T. Enhanced state-of-charge and state-of-health estimation of lithium-ion battery incorporating machine learning and swarm intelligence algorithm. *J. Energy Storage* **2024**, *83*, 110755. [CrossRef]
25. He, Y.; Bai, W.; Wang, L.; Wu, H.; Ding, M. SOH estimation for lithium-ion batteries: An improved GPR optimization method based on the developed feature extraction. *J. Energy Storage* **2024**, *83*, 110678. [CrossRef]
26. Buchanan, S.; Crawford, C. Probabilistic lithium-ion battery state-of-health prediction using convolutional neural networks and Gaussian process regression. *J. Energy Storage* **2024**, *76*, 109799. [CrossRef]
27. Dong, G.; Shen, F.; Sun, L.; Zhang, M.; Wei, J. A Bayesian Inferred Health Prognosis and State of Charge Estimation for Power Batteries. *IEEE Trans. Instrum. Meas.* **2025**, *74*, 1000312. [CrossRef]
28. Deng, Y.; Du, C.; Ren, Z. A Novel Method for Estimating the State of Health of Lithium-Ion Batteries Based on Physics-Informed Neural Network. *Batteries* **2025**, *11*, 49. [CrossRef]
29. Zhu, R.; Hu, J.; Peng, W. Bayesian calibrated physics-informed neural networks for second-life battery SOH estimation. *Reliab. Eng. Syst. Saf.* **2025**, *264*, 111432. [CrossRef]
30. Wang, N.; Xia, X.; Zeng, X. State of charge and state of health estimation strategies for lithium-ion batteries. *Int. J. Low-Carbon Technol.* **2023**, *18*, 443–448. [CrossRef]
31. Mulpuri, S.K.; Sah, B.; Kumar, P. Unraveling capacity fading in lithium-ion batteries using advanced cyclic tests: A real-world approach. *iScience* **2023**, *26*, 107770. [CrossRef]
32. Díaz, C.; Quintero, V.; Pérez, A.; Jaramillo, F.; Burgos-Mellado, C.; Rozas, H.; Orchard, M.E.; Sáez, D.; Cárdenas, R. Particle-Filtering-Based Prognostics for the State of Maximum Power Available in Lithium-Ion Batteries at Electromobility Applications. *IEEE Trans. Veh. Technol.* **2020**, *69*, 7187–7200. [CrossRef]
33. Zhang, C.; Jiang, J.; Zhang, L.; Liu, S.; Wang, L.; Loh, P.C. A Generalized SOC-OCV Model for Lithium-Ion Batteries and the SOC Estimation for LNMCO Battery. *Energies* **2016**, *9*, 900. [CrossRef]
34. Pozzato, G.; Allam, A.; Onori, S. Lithium-ion battery aging dataset based on electric vehicle real-driving profiles. *Data Brief* **2022**, *41*, 107995. [CrossRef]
35. Bustos, J.E.G.; Baeza, C.; Schiele, B.B.; Rivera, V.; Masserano, B.; Orchard, M.E.; Burgos-Mellado, C.; Perez, A. A novel data-driven framework for driving range prognostics in electric vehicles. *Eng. Appl. Artif. Intell.* **2025**, *142*, 109925. [CrossRef]
36. Pola, D.A.; Navarrete, H.F.; Orchard, M.E.; Rabié, R.S.; Cerda, M.A.; Olivares, B.E.; Silva, J.F.; Espinoza, P.A.; Pérez, A. Particle-Filtering-Based Discharge Time Prognosis for Lithium-Ion Batteries With a Statistical Characterization of Use Profiles. *IEEE Trans. Reliab.* **2015**, *64*, 710–720. [CrossRef]
37. Perez, A.; Jaramillo, F.; Baeza, C.; Valderrama, M.; Quintero, V.; Orchard, M. A Particle-Swarm-Optimization-Based Approach for the State-of-Charge Estimation of an Electric Vehicle When Driven Under Real Conditions. In Proceedings of the Annual Conference of the Prognostics and Health Management Society, Virtual, 29 November–2 December 2021; Volume 13. [CrossRef]
38. Patrizi, G.; Canzanella, F.; Ciani, L.; Catelani, M. Towards a State of Health Definition of Lithium Batteries through Electrochemical Impedance Spectroscopy. *Electronics* **2024**, *13*, 1438. [CrossRef]
39. Yao, L.; Xu, S.; Lin, X.; Li, X.; Zhang, Y. A Review of Lithium-Ion Battery State of Health Estimation and Prediction Methods. *World Electr. Veh. J.* **2021**, *12*, 113. [CrossRef]
40. Saha, B.; Goebel, K. Modeling Li-ion battery capacity depletion in a particle filtering framework. In Proceedings of the Annual Conference of the PHM Society, San Diego, CA, USA, 27 September–1 October 2009; Volume 1, pp. 11–20.
41. O'Kane, S.E.; Ai, W.; Madabattula, G.; Alonso-Alvarez, D.; Timms, R.; Sulzer, V.; Edge, J.S.; Wu, B.; Offer, G.J.; Marinescu, M. Lithium-ion battery degradation: How to model it. *Phys. Chem. Chem. Phys.* **2022**, *24*, 7909–7922. [CrossRef] [PubMed]
42. Zheng, Y.; Cui, Y.; Han, X.; Dai, H.; Ouyang, M. Lithium-ion battery capacity estimation based on open circuit voltage identification using the iteratively reweighted least squares at different aging levels. *J. Energy Storage* **2021**, *44*, 103487. [CrossRef]
43. Hu, X.L.; Ho, P.H.; Peng, L. Performance Analysis of Maximum Likelihood Estimation for Transmit Power Based on Signal Strength Model. *J. Sens. Actuator Netw.* **2018**, *7*, 38. [CrossRef]
44. Zhang, C.; Jin, M.; Dong, G.; Wei, S. Performance Analysis of the Maximum Likelihood Estimation of Signal Period Length and Its Application in Heart Rate Estimation with Reduced Respiratory Influence. *Appl. Sci.* **2023**, *13*, 10402. [CrossRef]
45. Rossi, R.J. *Mathematical Statistics: An Introduction to Likelihood Based Inference*; John Wiley & Sons: New York, NY, USA, 2018; p. 227. [CrossRef]
46. Yao, L.; Boughton, K.A. A Multidimensional Item Response Modeling Approach for Improving Subscale Proficiency Estimation and Classification. *Appl. Psychol. Meas.* **2007**, *31*, 83–105. [CrossRef]

47. Dore, A.; Pinasco, M.; Regazzoni, C.S. CHAPTER 9—Multi-Modal Data Fusion Techniques and Applications. In *Multi-Camera Networks*; Aghajan, H., Cavallaro, A., Eds.; Academic Press: Oxford, UK, 2009; pp. 213–237. [CrossRef]
48. Hastings, W.K. Monte Carlo Sampling Methods Using Markov Chains and Their Applications. *Biometrika* **1970**, *57*, 97–109. [CrossRef]
49. Zając, M. Online fault detection of a mobile robot with a parallelized particle filter. *Neurocomputing* **2014**, *126*, 151–165. [CrossRef]
50. Arulampalam, M.; Maskell, S.; Gordon, N.; Clapp, T. A tutorial on particle filters for online nonlinear/non-Gaussian Bayesian tracking. *IEEE Trans. Signal Process.* **2002**, *50*, 174–188. [CrossRef]
51. Paccha-Herrera, E.; Jaramillo-Montoya, F.; Calderón-Muñoz, W.R.; Tapia-Peralta, D.; Solórzano-Castillo, B.; Gómez-Peña, J.; Paccha-Herrera, J. A particle filter-based approach for real-time temperature estimation in a lithium-ion battery module during the cooling-down process. *J. Energy Storage* **2024**, *94*, 112413. [CrossRef]
52. Doucet, A.; Godsill, S.; Andrieu, C. On sequential Monte Carlo sampling methods for Bayesian filtering. *Stat. Comput.* **2000**, *10*, 197–208. [CrossRef]
53. Vella, A.; Alonso, M.A. Chapter Seven—Maximum likelihood estimation in the context of an optical measurement. In *A Tribute to Emil Wolf*; Visser, T.D., Ed.; Progress in Optics; Elsevier: Amsterdam, The Netherlands, 2020; Volume 65, pp. 231–311. [CrossRef]
54. Jiang, G.; Li, M.; Liu, Z.; Sun, L.; Wang, Q. A Maximum Likelihood Estimation Method for Underwater Radiated Noise Power. *Appl. Sci.* **2025**, *15*, 6692. [CrossRef]
55. Myung, I.J. Tutorial on maximum likelihood estimation. *J. Math. Psychol.* **2003**, *47*, 90–100. [CrossRef]
56. Doucet, A.; Freitas, N.; Gordon, N. *Sequential Monte Carlo Methods in Practice*, 1st ed.; Springer: New York, NY, USA, 2001. [CrossRef]
57. Cappé, O.; Godsill, S.; Moulines, E. An overview of existing methods and recent advances in sequential Monte Carlo. *Proc. IEEE* **2007**, *95*, 899–924. [CrossRef]
58. Burgos-Mellado, C.; Orchard, M.E.; Kazerani, M.; Cárdenas, R.; Sáez, D. Particle-filtering-based estimation of maximum available power state in Lithium-Ion batteries. *Appl. Energy* **2016**, *161*, 349–363. [CrossRef]
59. Zeng, Z.; Tian, J.; Li, D.; Tian, Y. An Online State of Charge Estimation Algorithm for Lithium-Ion Batteries Using an Improved Adaptive Cubature Kalman Filter. *Energies* **2018**, *11*, 59. [CrossRef]
60. Lee, S.; Kim, J.; Lee, J.; Cho, B. State-of-charge and capacity estimation of lithium-ion battery using a new open-circuit voltage versus state-of-charge relationship. *J. Power Sources* **2008**, *185*, 1367–1373. [CrossRef]
61. Olivares, B.E.; Cerda, M.A.; Orchard, M.E.; Silva, J.F. Particle-Filtering-Based Prognosis Framework for Energy Storage Devices with a Statistical Characterization of State-of-Health Regeneration Phenomena. *IEEE Trans. Instrum. Meas.* **2013**, *62*, 364–376. [CrossRef]
62. Tampier, C.; Pérez, A.; Jaramillo, F.; Quintero, V.; Orchard, M.E.; Silva, J.F. Lithium-Ion Battery End-of-Discharge Time Estimation and Prognosis based on Bayesian Algorithms and Outer Feedback Correction Loops: A Comparative Analysis. In Proceedings of the Annual Conference of the Prognostics and Health Management Society 2015, Austin, TX, USA, 18–24 October 2015; Volume 7. [CrossRef]
63. Brito Schiele, B. Health Inference and Diagnostic Architecture Based on Bayesian Filtering and Maximum Likelihood Estimation for Electromobility and Structural Engineering. Master's Thesis, Universidad de Chile, Santiago, Chile, 2024. Available online: https://repositorio.uchile.cl/handle/2250/203347 (accessed on 16 November 2025).
64. Baek, S.; Choi, W. Performance Comparison of Machine Learning-Based Static Capacity Estimation Technique with Various C-rate Partial Discharges. In Proceedings of the 2024 IEEE Transportation Electrification Conference and Expo, Asia-Pacific (ITEC Asia-Pacific), Xi'an, China, 10–13 October 2024; pp. 228–233. [CrossRef]
65. Ding, X.; Zhang, D.; Cheng, J.; Wang, B.; Luk, P.C.K. An improved Thevenin model of lithium-ion battery with high accuracy for electric vehicles. *Appl. Energy* **2019**, *254*, 113615. [CrossRef]

MDPI AG
Grosspeteranlage 5
4052 Basel
Switzerland
Tel.: +41 61 683 77 34

Batteries Editorial Office
E-mail: batteries@mdpi.com
www.mdpi.com/journal/batteries

www.ingramcontent.com/pod-product-compliance
Lightning Source LLC
Chambersburg PA
CBHW041949260926
51850CB00019B/2091
9783725863327